Elements

IRON
CHROMIUM AND MANGANESE

Fe

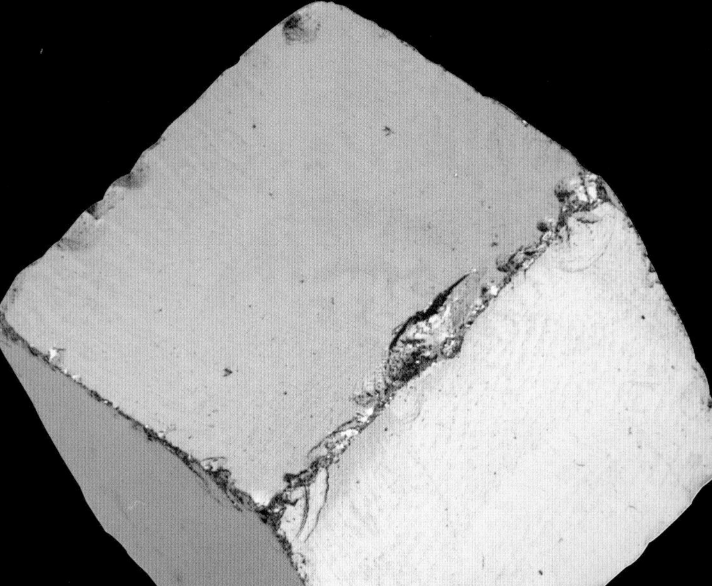

Cr

Mn

Atlantic Europe Publishing

How to use this book

This book has been carefully developed to help you understand the chemistry of the elements. In it you will find a systematic and comprehensive coverage of the basic qualities of each element. Each two-page entry contains information at various levels of technical content and language, along with definitions of useful technical terms, as shown in the thumbnail diagram to the right. There is a comprehensive glossary of technical terms at the back of the book, along with an extensive index, key facts, an explanation of the Periodic Table, and a description of how to interpret chemical equations.

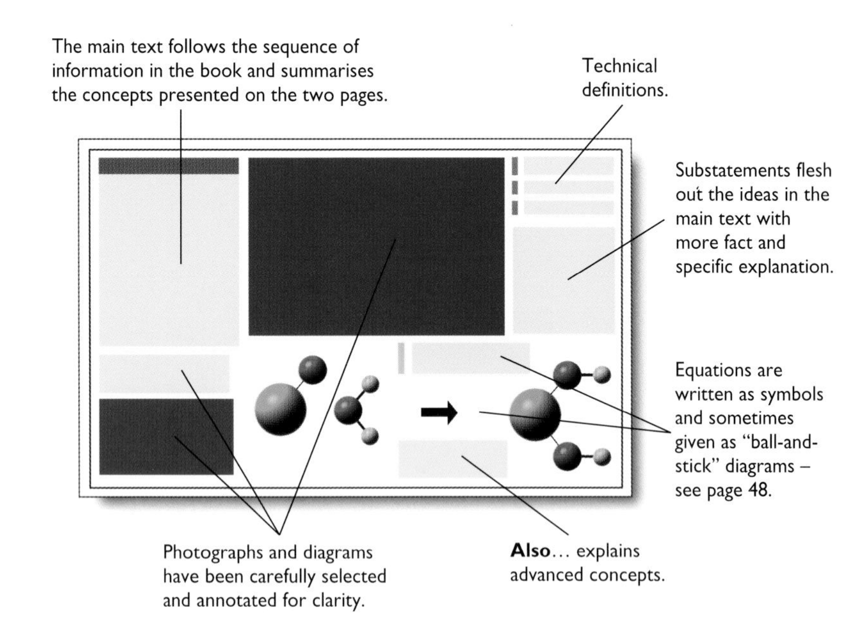

An Atlantic Europe Publishing Book

Author
Brian Knapp, BSc, PhD
Project consultant
Keith B. Walshaw, MA, BSc, DPhil
(Head of Chemistry, Leighton Park School)
Industrial consultant
Jack Brettle, BSc, PhD (Chief Research Scientist, Pilkington plc)
Art Director
Duncan McCrae, BSc
Editor
Elizabeth Walker, BA
Special photography
Ian Gledhill
Illustrations
David Woodroffe
Electronic page make-up
Julie James Graphic Design
Designed and produced by
EARTHSCAPE EDITIONS
Print consultants
Landmark Production Consultants Ltd
Reproduced by
Leo Reprographics
Printed and bound by
Paramount Printing Company Ltd

First published in 1996 by
Atlantic Europe Publishing Company Limited, Greys Court Farm,
Greys Court, Henley-on-Thames, Oxon, RG9 4PG, UK.

Suggested cataloguing location
Knapp, Brian
Iron, chromium and manganese
ISBN 1 869860 24 1
– *Elements* series
540

Acknowledgements
The publishers would like to thank the following for their kind help and advice: *Mr Ah Aia Tan and Guthrie Plantation & Agricultural Sdn Bhd, Mr J. Made Mawa of Budi-Ukir, Mr I. Made Rangun, Pippa Trounce* and *Vauxhall Motors Limited.*

Picture credits
All photographs are from the **Earthscape Editions** photolibrary except the following:
(c=centre t=top b=bottom l=left r=right)
Vauxhall Motors Limited 11t, 33b and **ZEFA** 31cl.

Front cover: The chemical reduction of iron oxide to molten iron. At the time this picture was taken the mixture had reached 2000°C.
Title page: A cubic crystal of pyrite or iron sulphide.

Contents

Introduction

An element is a substance that cannot be broken down into a simpler substance by any known means. Each of the 92 naturally occurring elements is therefore one of the fundamental materials from which everything in the Universe is made. This book is about iron, chromium and manganese.

Iron

Iron is one of the world's workhorse elements, found in most of the structures we make, from bridges and skyscrapers to computers and fencing wire.

Iron is cheap to obtain, easy to shape and very strong. For all these reasons it is used more widely than any other metal. But all of the advantages of iron have to be balanced against one major disadvantage: it is a fairly reactive element, prone to rust when exposed to damp air.

Iron (whose chemical symbol, Fe, comes from the Latin *ferrum*) is now one of the most commonly used metals of modern times. But it was not always so. Although the first use of iron dates back some three thousand years to the period of archaeology called the Iron Age, until the last century iron was difficult to find and work and expensive to use.

Pure iron is a soft silvery-grey metal. It can be bent and stretched at room temperature, and at 1535°C it will melt. This temperature is much higher than the temperature at which wood burns. Earlier civilisations found iron so difficult to use because they could not produce such high temperatures to work iron.

In fact iron has only been widely available since the Industrial Revolution of the 18th and 19th centuries. At this time, inventors like Abraham Darby learned how to obtain large quantities of iron economically using coke. With iron at last cheap and plentiful, the people of the 19th century led the world into the engineering age.

Iron is not just found as a metal. Compounds of iron, for example, are found in almost all living things: iron compounds are a vital nutrient for all plants and animals, and they make our blood appear red in colour. Iron compounds are also the basis of a rich variety of natural colours, both in rocks and in nature.

Iron, along with very few other elements, possesses the property of magnetism. This property makes iron essential in compasses and inside every electric motor. Indeed, this makes iron one of the most versatile of all the elements.

Chromium

Chromium, whose chemical symbol is Cr, is named after the Greek word for colour. It is responsible for both the deep red colour of a ruby and the green of emeralds. Chromium is a shiny, rare metal, but because it resists corrosion, it is very important as a surface coating called "chromium plating".

▲ Iron is the colouring in the red haemoglobin molecules of blood cells. Its main function is to carry oxygen around the body.

Foods high in iron content include meat (especially liver and heart), egg yolk, wheat germ, and most green vegetables.

Some people, such as pregnant women and elderly people, suffer iron deficiency. They have to take extra iron in the form of tablets (like those shown here).

Manganese

Manganese, whose chemical symbol is Mn, is named for the Latin word for magnesia, a magnetic stone. It is a silver–grey metal. Manganese has been identified in great quantities on the deep ocean floor, as yet too deep for collection. Manganese is used for hardening steel and in batteries.

Iron

Iron is a dense, silvery-grey metal. It is the most widely used of all the metal elements because it is common, is cheap to produce from its ores, can be bent while cold, and is strong.

However, some of its properties are less welcome. For example, it is quite reactive, especially with air and water, a process called corrosion. You can see why this happens by looking at the demonstrations below.

❶ Iron filings (the dark material) are placed in a small pile in a dish. Copper sulphate (the blue solution) is added.

The reactivity of iron

Some metals are more reactive than others. Gold, for example, hardly reacts at all, which is why it stays bright and does not tarnish (develop a dull oxide coating). On the other hand, some metals, like potassium, react vigorously with oxygen (corrode) as soon as they are placed in the air.

Metals can be placed in order of how vigorously they react, in a list called a reactivity series (see right).

Copper is near the bottom of the reactivity series, while iron is higher up. Therefore iron will always corrode when placed in a solution containing a copper compound such as copper sulphate. On the other hand, iron is below magnesium, which is why magnesium will always corrode when placed next to iron. This is shown in the picture on the right, where a small strip of magnesium has been wrapped around the shank of an iron nail and placed in a bottle containing water and an indicator.

REACTIVITY SERIES	
Element	*Reactivity*
potassium	*most reactive*
sodium	
calcium	
magnesium	
aluminium	
manganese	
chromium	
zinc	
iron	
cadmium	
tin	
lead	
copper	
mercury	
silver	
gold	
platinum	*least reactive*

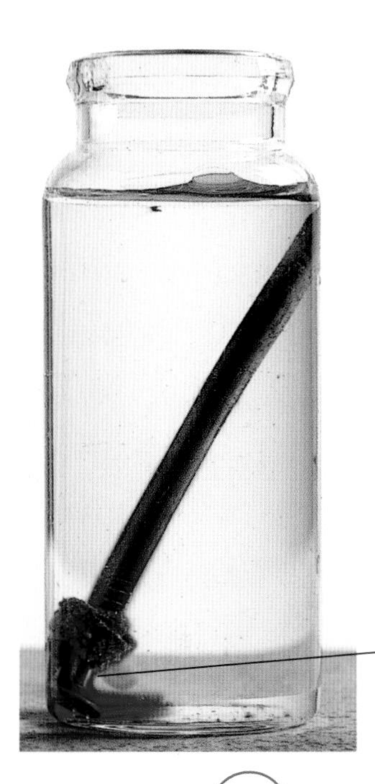

The red colour is caused by an indicator in the water. The clear water shows a neutral reaction. As the magnesium corrodes it produces magnesium hydroxide, an alkali, which turns the region near the corroding magnesium red.

❷ After a while a number of changes have occurred because the iron filings have reacted with the copper sulphate. There is no acid or other chemical here.

Notice how the solution has become a very pale green. The edges of the iron filings have changed to a reddish-brown (coppery) colour, looking a little like a reef fringing a coral island.

The reason for the changes is that a chemical reaction has occurred. Some of the iron has gone into solution. The copper "reef" has been deposited (precipitated) out of the solution onto the iron filings.

The copper in the copper sulphate makes the solution blue; iron sulphate solutions are green, so the green colour shows that the copper and iron have "swapped", and the solution is now iron sulphate.

However, not all of the iron has reacted. There is far more iron than copper, so a complete swap cannot be achieved. If the iron were placed in a huge vat of copper sulphate it would eventually react entirely and disappear.

cathodic protection: the technique of making the object that is to be protected from corrosion into the cathode of a cell. For example, a material, such as steel, is protected by coupling it with a more reactive metal, such as magnesium. Steel forms the cathode and magnesium the anode. Zinc protects steel in the same way.

corrosion: the *slow* decay of a substance resulting from contact with gases and liquids in the environment. The term is often applied to metals. Rust is the corrosion of iron.

reactivity: the tendency of a substance to react with other substances. The term is most widely used in comparing the reactivity of metals. Metals are arranged in a reactivity series.

solution: a mixture of a liquid and at least one other substance (e.g. salt water). Mixtures can be separated out by physical means, for example by evaporation and cooling.

Also...

This demonstration uses iron filings, the kind of "swarf" that might be produced as waste from a metal lathe. If a large mass of iron is present as iron filings, there is a larger surface area than if it were present in a large block. As a result, chemical reactions will happen more quickly.

EQUATION: Reaction between iron and copper sulphate

Iron + copper sulphate ⇨ iron sulphate + copper

$Fe(s) + CuSO_4(aq) \Rightarrow FeSO_4(aq) + Cu(s)$

Corrosion of iron: rust

The environment is a hazardous place for materials. Even in "clean" country air, materials will slowly show signs of surface change because of the effects of water and gases in the air. The change in the surface of materials is a reaction with oxygen in the air. It is called corrosion and produces a coating called an oxide.

Iron is particularly susceptible to corrosion, known as rusting, in damp air or oxygen-rich water because it is a reactive material.

▼ This jar is called a desiccator. The blue crystals in the bottom are silica gel coloured with cobalt chloride. They absorb any moisture in the air. The nails are therefore in completely dry air. Although they have been sealed in this desiccator for many years, they show no signs of rust. This indicates that both moisture and oxygen are required to form rust.

▲ Steel nails need water and oxygen from the air to rust quickly. Water alone is a poor corroding agent. However, when oxygen is present, such as when water is left uncovered, nails left in water will rust (oxidise) very rapidly.

Iron's special oxide layers

The materials that most often show signs of corrosion are metals. The oxide coating that develops on the surface of some metals is so thin it is invisible.

Look at a clean iron or steel nail and the surface looks unaffected because the oxide layer is so thin. When the oxide coating is thicker, it may appear as a discolouring, or tarnishing, of the surface.

However, unlike some other metals such as aluminium and copper, iron's oxide coating is not able to keep water and oxygen out. On the contrary, it is a porous coating, which is why it rusts.

corrosion: the *slow* decay of a substance resulting from contact with gases and liquids in the environment. The term is often applied to metals. Rust is the corrosion of iron.

oxide: a compound that includes oxygen and one other element.

porous: a material containing many small holes or cracks. Quite often the pores are connected, and liquids, such as water or oil, can move through them.

product: a substance produced by a chemical reaction.

◀ Old iron anchoring chain and a new galvanised chain.

▼ The anchor chain is not attacked evenly. The surface of the iron is pitted in places and shows rust scales in other places.

▼ This rusty bolt shows many of the features of corrosion. Notice, for example, how the thread is much less clear. This is because when iron corrodes (rusts), oxygen and iron combine to make a bulkier substance than pure iron. So, as the sides of the threads rust and swell, the gap between threads becomes smaller. This is one reason why a rusty bolt is so hard to undo: the oxide has jammed the threads.

The nature of rust

When a material rusts it is because water and oxygen have combined with the metal iron. To understand what happens, chemists think about how the tiniest particles – the atoms – of each element behave when they react.

The unrusty iron is made of small heavy atoms that pack tightly together. The water is a molecule made of one oxygen atom and two hydrogen atoms. The oxygen gas is made up of pairs of oxygen atoms.

When rusting occurs on the surface of a material, the atoms split up and regroup. The combination of several different sizes of atoms is a bulkier and less dense material than pure iron, and it easily flakes off.

▼ Large structures like the Sydney Harbour bridge need to be painted constantly to prevent them from corroding away. Even so, corrosion in exposed places is difficult to prevent.

Rust and its prevention

Iron or mild steel, water and the oxygen in air combine to cause rust. It is one of the most common chemical reactions around us. Rusting occurs even in damp air – no actual droplets of water are needed – and even more quickly in air that has impurities in it. Impurities can include sea salt (which is one reason ships and iron objects near the sea rust so easily) and also pollution gases (such as may be found in cities and near factories or power stations). Sulphur dioxide (present in acid rain) is a particularly effective pollutant because it dissolves in water droplets to form sulphuric acid.

We use iron and steel in all manner of objects because it is cheap, strong and plentiful. To counteract rusting, iron is protected with paint or another metal such as zinc (when it is called galvanised iron).

▼ All iron and steel objects have to be painted or protected in some other way against rust. Most vehicles are coated in a number of protective layers during manufacture and given several coats of paint before they leave the factory.

electrode: a conductor that forms one terminal of a cell.

electrolysis: an electrical–chemical process that uses an electric current to cause the break up of a compound and the movement of metal ions in a solution. The process happens in many natural situations (as for example in rusting) and is also commonly used in industry for purifying (refining) metals or for plating metal objects with a fine, even metal coating.

electrolyte: a solution that conducts electricity.

oxidation/reduction: a reaction in which oxygen is gained or lost. (Also... More generally oxidation involves the loss of electrons.)

EQUATION: The rusting of iron

Iron + water + oxygen ➪ ferric hydroxide ➪ ferric oxide + water

$$4Fe(s) + 6H_2O(l) + 3O_2(g) \Rightarrow 4Fe(OH)_3(s) \Rightarrow \underbrace{2Fe_2O_3(s)}_{\text{Rust}} + 6H_2O(l)$$

Also...
How iron becomes pitted

If a wetted iron surface is exposed either by being uncoated or because the paint on the surface has been chipped, oxygen atoms are able to enter the water through its surface skin. In this water one of the world's tiniest batteries forms. The water is oxygen rich and the iron forms an electrode, one terminal of a battery. The oxygen-poor region in the scratch, and thus farther from the air, forms the other electrode. The water forms the electrolyte. A minute electric current now flows, and iron is carried in solution to the oxygen-rich water, where it is oxidised and deposited.

Thus, iron is removed from one part of the metal and deposited as an oxide or hydroxide nearby. This explains why rusty material is often both pitted and lumpy.

(To find out more about the way electrolysis and electroplating work, see page 37.)

▶ This rusting horseshoe shows two forms of rust. The light brown rust patches are recently formed ferric hydroxide or $Fe(OH)_3$. In contrast, the darker-brown rust patches are the final solid state of ferric oxide, or Fe_2O_3, shown in the equation above.

Ores of iron

Iron is the second most abundant metal (after aluminium) in the Earth's crust. However, pure iron is seldom found in nature. Most often it is found in the ore called haematite (which contains one-third of its weight as iron) and magnetite (with about two-thirds of its weight as iron). These compounds, both containing iron and oxygen, are known as iron oxides.

Magnetite

Iron is famous for its ability to act as a magnet, or to be attracted by magnets. Native (pure) iron as well as iron alloys such as steel, and iron compounds such as some iron oxides (for example the ore magnetite) are also magnetic. The property is created because each tiny crystal of iron can behave as a magnet, organising itself into the same direction as all those crystals nearby.

The earliest knowledge of the magnetic properties of iron comes from the strongly magnetic rock known as lodestone or magnetite. The word magnetite comes from the region called Magnesia in Greece, where lodestone was mined in ancient times.

A ball-shaped piece of lodestone has two regions where it will attract or repel other ball-shaped lodestones. These places are known as the magnetic poles.

▶ This piece of magnetite has been dipped in iron filings to demonstrate its magnetic properties. Magnetite is Fe_3O_4.

Meteorites

Meteorites are one of the few highly concentrated sources of iron ore. In some meteorites iron occurs as uncombined metal (called native metal).

Because the majority of all meteorites are about nine-tenths iron, they require relatively little purifying and for this reason they were prized by earlier civilisations. Huge craters where meteorites had fallen were an obvious site to find native iron.

In ancient times, iron was known as the "metal of heaven". This may be because people saw meteorites fall and then discovered that they contained almost pure iron. The first iron used was definitely from meteorites like the one that created the giant Meteor Crater in Arizona.

native metal: a pure form of a metal, not combined as a compound. Native metal is more common in poorly reactive elements than in those that are very reactive.

ore: a rock containing enough of a useful substance to make mining it worthwhile.

oxidation: a reaction in which the oxidising agent removes electrons. (Note that oxidising agents do not have to contain oxygen.)

▲ A piece of iron ore (haematite). The deep red colour is due to the fact that haematite is iron oxide. Pure iron forms the grey coloured crystals. Haematite is Fe_2O_3.

Haematite

This is the name for the most widespread form of iron ore. It is most often found in rocks once deposited by rivers or the sea.

The deep red colour of some rocks indicates that they contain haematite. Most of these rocks were formed in parts of the tropics with wet and dry seasons. During the wet season the minerals eroded from the land were washed to basins, deltas or coasts. During the dry period the water evaporated and the sediments dried out, and iron compounds oxidised to iron oxide. These rocks are often referred to as "red beds" and are usually fine-grained materials such as shales.

Limonite

Limonite is an iron oxide that forms in cooler climates than haematite. It is usually yellow, orange and brown, rather than red. It contains water molecules, and so is an example of a hydrous (water-containing) iron oxide.

Limonite is often found in marshes and is thus also known as bog iron ore.

The water content influences the colour of the oxide. Heating limonite causes the water to be driven off, thus darkening the colour of the oxide and yielding the paint colour "burnt ochre". In fact, it was traditionally used as a source of ochre pigments in paint.

Iron oxides

Iron oxide can occur in two forms, depending on how much oxygen is bound with the iron. Compounds of iron with a lower oxygen content are known as ferrous compounds, whereas the compounds of iron with the higher oxygen content are known as ferric compounds.

Ferrous compounds

The colour of the iron compound is related to the amount of oxygen it contains. Ferrous compounds (also known as iron II compounds) have a green, grey or blue colour.

The dirty-looking green gelatinous precipitate in the bottom of the tube on the right is a ferrous hydroxide (iron II hydroxide).

Ferric compounds

Ferric compounds (also known as iron III compounds) have a yellow, red or brown colour.

The red–brown gelatinous precipitate in the bottom of the tube shown here is ferric hydroxide (iron III hydroxide).

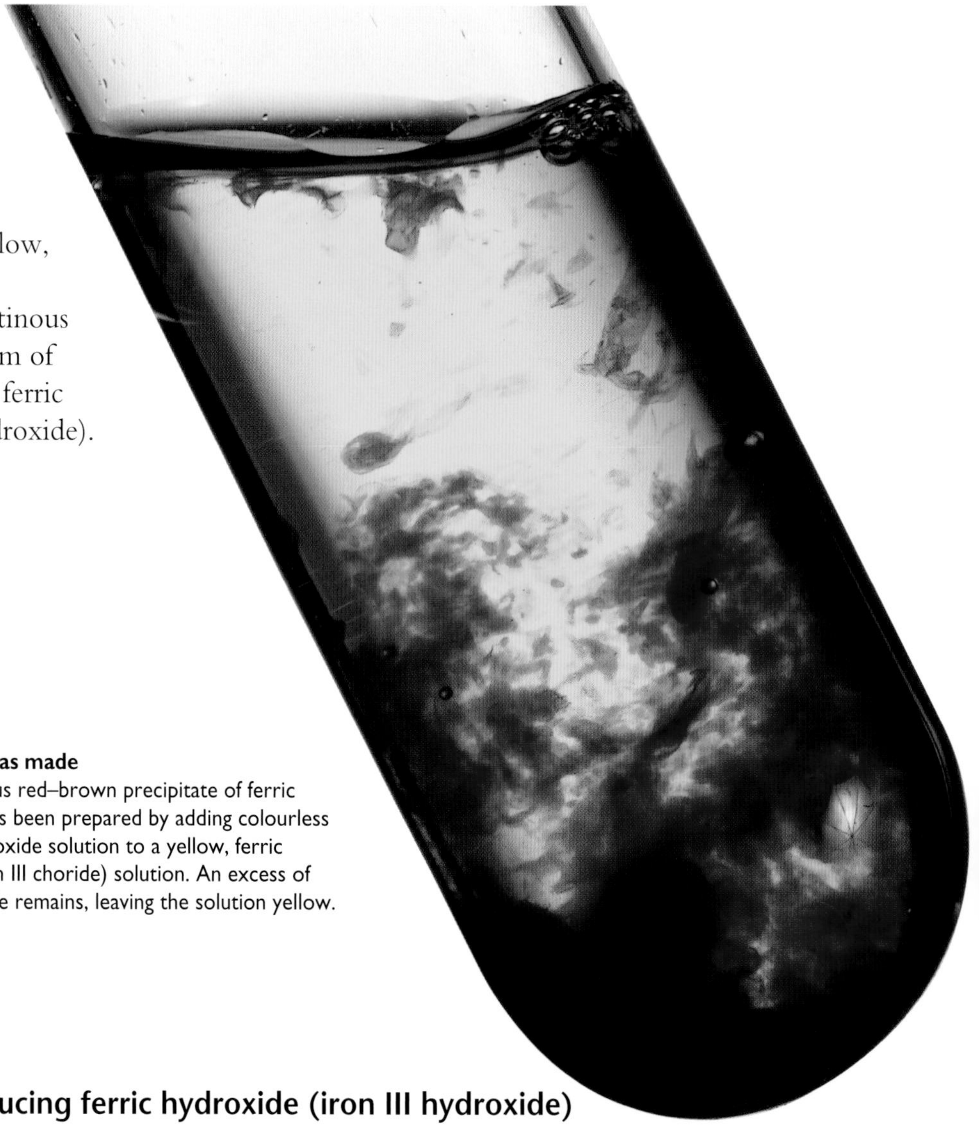

▶ How it was made
The gelatinous red–brown precipitate of ferric hydroxide has been prepared by adding colourless sodium hydroxide solution to a yellow, ferric chloride (iron III choride) solution. An excess of ferric chloride remains, leaving the solution yellow.

EQUATION: Producing ferric hydroxide (iron III hydroxide)

Sodium hydroxide + iron III chloride ⇨ iron III hydroxide + sodium chloride

3NaOHaq) + $FeCl_3$(aq) ⇨ $Fe(OH)_3$(s) + 3NaCl(aq)

gelatinous: a term meaning made with water. Because a gelatinous precipitate is mostly water, it is of a similar density to water and will float or lie suspended in the liquid.

Also...

Many other iron compounds have striking colours. They have been known and used since ancient times.

Ochre

Ochre is the name given to yellow–red colours produced by iron compounds. Ochre is a powder made by crushing many iron-rich ores. Limonite and haematite (see page 13) produce yellow and red colouring, respectively. If either of these ores is roasted, the colours will darken to red–brown. This is called burnt ochre. The powder is used in paints as the colouring agent or pigment. It is mixed with a liquid before being applied.

Prussian blue

Prussian blue (so named because it was developed in Prussia during the 18th century) is obtained by reacting iron oxide with potassium ferrocyanide. It is used as a "whitener" in washing powders, being a light blue dye that gives the optical illusion of whiteness to clothes. In more concentrated form it acts as a blue pigment in paints and enamels.

▶ How it was made

The green, gelatinous precipitate of ferrous hydroxide (iron II hydroxide) was prepared by adding colourless sodium hydroxide solution to ferrous sulphate (iron II sulphate) solution.

EQUATION: Producing ferrous hydroxide (iron II hydroxide)

Sodium hydroxide + iron II chloride ➪ iron II hydroxide + sodium chloride

$$2NaOH(aq) + FeCl_2(aq) \Rightarrow Fe(OH)_2(s) + 2NaCl(aq)$$

Iron colours in the landscape

Much of the colour in the world's rocks and soils is produced by iron compounds. These colours allow geologists and soil scientists to tell much about the way the rocks and soils of the world have formed.

Deep red colours tell of iron that was oxidised in hot, tropical conditions; oranges and yellows tell of iron oxides formed in cooler climates; while greys, blues and greens are produced where iron compounds were short of oxygen, such as deep under the sea.

Soils formed from these rocks often have a colour similar to that of the rock on which it has formed. However, when soils are affected by acid waters, iron is removed into solution, and the rock becomes dramatically paler in colour.

Iron and soils

Iron compounds are important constituents of soils and responsible for many soil colours. In some parts of the world they are also responsible for poor drainage and a variety of other soil problems.

Soils are coloured by a combination of iron oxides and organic matter. Most organic matter occurs close to the surface. Its black colour makes the topsoil a darker colour than any other part of the soil.

In some cases, such as when a soil receives large amounts of rainfall, the organic matter reacts with the rainwater to produce acids that are strong enough to dissolve the iron compounds. When this happens, the iron compounds, which provide the brown, orange, red and yellow colours in a soil, are removed and the soil layer turns ashen grey. The ashen grey colour is the colour of the quartz sand and clay particles that remain.

In regions lower down in the soil, the acid conditions are less intense and the iron compounds are reprecipitated. Here the iron concentrates to produce a strong orange colour.

▲ Soils change colour through their profiles, in part because iron is carried by acid waters, as shown in this Podzol from New Zealand.

acidity: a general term for the strength of an acid in a solution.

oxide: a compound that includes oxygen and one other element.

solution: a mixture of a liquid and at least one other substance (e.g. salt water). Mixtures can be separated out by physical means, for example by evaporation and cooling.

◀ The reddish-coloured banding of this sandstone shows that these rocks were formed in hot conditions. The variation in colour shows that some materials were formed in more iron-rich environments than others. This is Delicate Arch, Arches National Park, Utah, USA.

Laterite

Most tropical soils have a deep red colour, which is the result of weathering of iron compounds in climates with high temperatures.

Laterite is the name given to a soil material that is found in some tropical regions with a long dry season followed by long wet season.

A laterite layer is a thick zone containing iron and aluminium oxides and very little else. This unusual material is so rich in iron compounds that it is mined as an iron ore.

Most laterites are very old. It is believed that they result from the effect of large amounts of water washing through a tropical soil. Under these conditions the silica that makes up the body of sand and clay particles (normally not a soluble material) goes into solution, so that the iron and aluminium compounds remain behind and so become more concentrated.

While still in the soil, laterite layers are soft, but when they are exposed to the air, such as when a road cutting is made, they oxidise and quickly become rock hard.

◀ This laterite soil from Queensland in Australia is very deep and on exposure to air becomes very hard.

Iron sulphide

Iron sulphide, or pyrite, is a common compound of iron, also known as "fool's gold". Pyrite is a very common mineral, occurring in many types of rock. It is commonly found in veins and can form perfect cubic (cube-shaped) crystals.

Shale rocks contain pyrite because the muds from which they are formed were once also combined with decaying organic matter. As the organic matter decayed, it took oxygen from the water, and hydrogen sulphide gas was produced. The hydrogen sulphide reacted with iron in the water to produce iron sulphide. In the right conditions the iron sulphide grew into perfect cubic crystals; in other circumstances it developed into nodules.

Iron sulphide is not stable in damp air, where it readily oxidises to a brown colour. The shiny crystal that you see in the picture on the right will turn progressively grey and begin to crumble to iron sulphate.

Iron sulphide is not used as an iron ore, because its high sulphur content is difficult to remove. It is mainly used in the production of sulphuric acid.

A cubic crystal of pyrite. The parallel "scratches", known as striations, are a common feature of pyrite crystals.

▶ This is a typical sample of an ore. The conditions that allow the formation of pyrite are also right for the formation of crystals of other compounds. This is the reason refining ores is often a very complicated chemical process.

These transparent crystals are quartz.

The dark crystals are sphalerite (zinc sulphide).

ion: an atom, or group of atoms, that has gained or lost one or more electrons and so developed an electrical charge.

pyrite: "mineral of fire". This name comes from the fact that pyrite (iron sulphide) will give off sparks if struck with a stone.

sulphide: a sulphur compound that contains no oxygen.

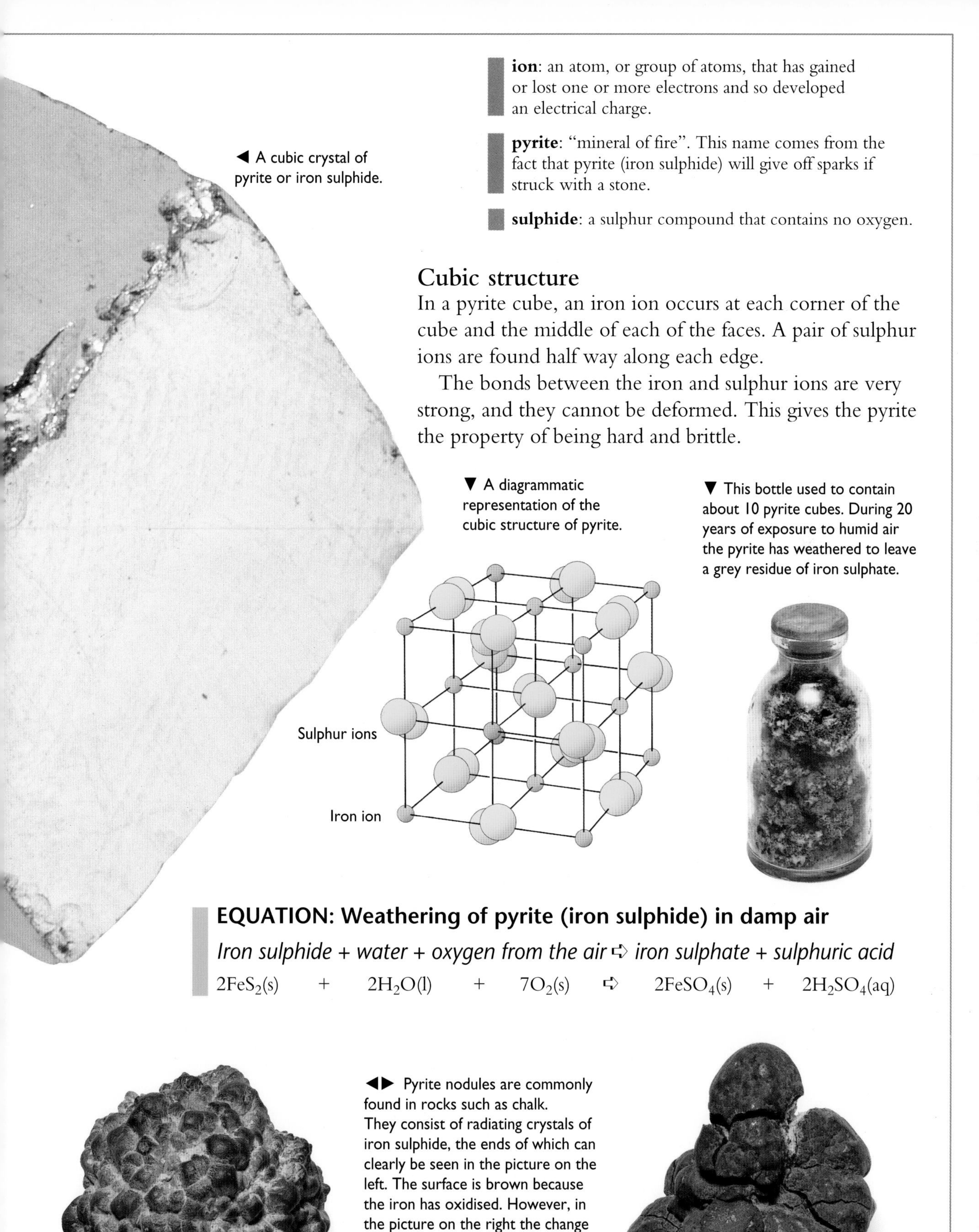

◀ A cubic crystal of pyrite or iron sulphide.

Cubic structure

In a pyrite cube, an iron ion occurs at each corner of the cube and the middle of each of the faces. A pair of sulphur ions are found half way along each edge.

The bonds between the iron and sulphur ions are very strong, and they cannot be deformed. This gives the pyrite the property of being hard and brittle.

▼ A diagrammatic representation of the cubic structure of pyrite.

▼ This bottle used to contain about 10 pyrite cubes. During 20 years of exposure to humid air the pyrite has weathered to leave a grey residue of iron sulphate.

EQUATION: Weathering of pyrite (iron sulphide) in damp air

Iron sulphide + water + oxygen from the air ➪ iron sulphate + sulphuric acid

$2FeS_2(s) + 2H_2O(l) + 7O_2(s) \Rightarrow 2FeSO_4(s) + 2H_2SO_4(aq)$

◀▶ Pyrite nodules are commonly found in rocks such as chalk. They consist of radiating crystals of iron sulphide, the ends of which can clearly be seen in the picture on the left. The surface is brown because the iron has oxidised. However, in the picture on the right the change from sulphide to sulphate inside the nodule is causing it to break up.

Iron compounds

Iron is well known for its magnetic qualities, its ability to bend under pressure and its tendency to go rusty in air. But just how many, if any, of these characteristics are inherited by iron compounds?

The relationship of iron compounds to iron can be investigated by mixing iron and sulphur. The results of heating the mixture are described here.

❶▼ Iron and sulphur form distinctively coloured powders. When mixed the dark brown iron can be seen speckling the yellow sulphur.

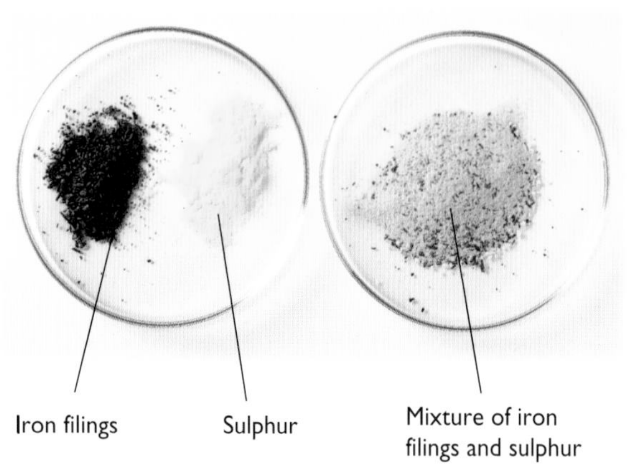

❷▲ Iron is magnetic, so a magnet will draw the iron from the iron/sulphur mixture. At room temperature the mixture can thus easily be separated.

❸▲ If the mixture is heated strongly with a Bunsen flame, as shown above, it will begin to glow red.

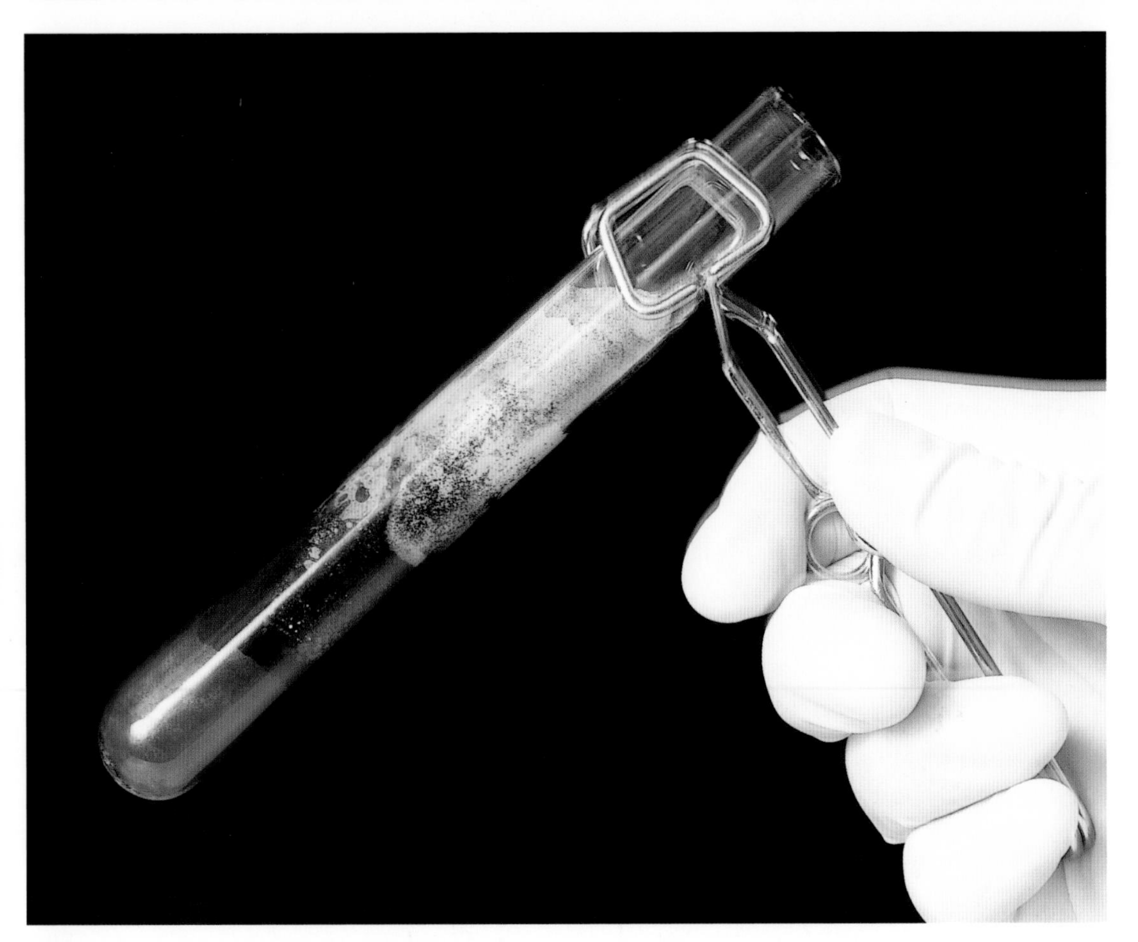

❹▲ The tube can then be taken from the flame and the red glow will spread up the tube without further heating. This is because the chemical reaction taking place, caused by heating, itself releases heat that in turn keeps the reaction going as long as there is still elemental iron and sulphur in the tube.

❺► When the reaction is over, there is a brown coke-like compound in the tube. This is iron sulphide. Compare it with the crystals on pages 18 and 19 to see that compounds can have the same chemical formula but look remarkably different. Iron sulphide has no magnetic properties, showing that the compound shares little in common with the starting materials from which it was made.

When heated in a test tube the iron and sulphur combine to form the compound iron sulphide, which has no magnetic properties.

EQUATION: Laboratory production of iron sulphide

Iron + sulphur ⇨ iron sulphide

Fe(s) + S(s) ⇨ FeS(s)

Refining iron ore: a small-scale demonstration

Iron oxide will not react at room temperature. This is why a piece of iron oxide rock will simply sit on your window ledge for ever.

In common with many chemical reactions, you need to put in a lot of heat energy to make iron oxide react with other chemicals.

In the demonstration shown on this page, the principle of generating heat by chemical reactions is shown.

▲ The magnesium ribbon fuse is lighted.

The apparatus

A container (in this case a plant pot with a hole in the bottom, because it is cheap and it only melts at very high temperatures) is filled with a mixture of aluminium powder and iron oxide (iron ore) powder. A small cup-shaped indentation is made in the top of the mixture and some barium peroxide powder is added. A fuse of magnesium ribbon is buried in the top of the mixture.

> ***WARNING***
> *This demonstration produces materials with a temperature of 2000°C and must never be attempted by anyone except a qualified chemist.*

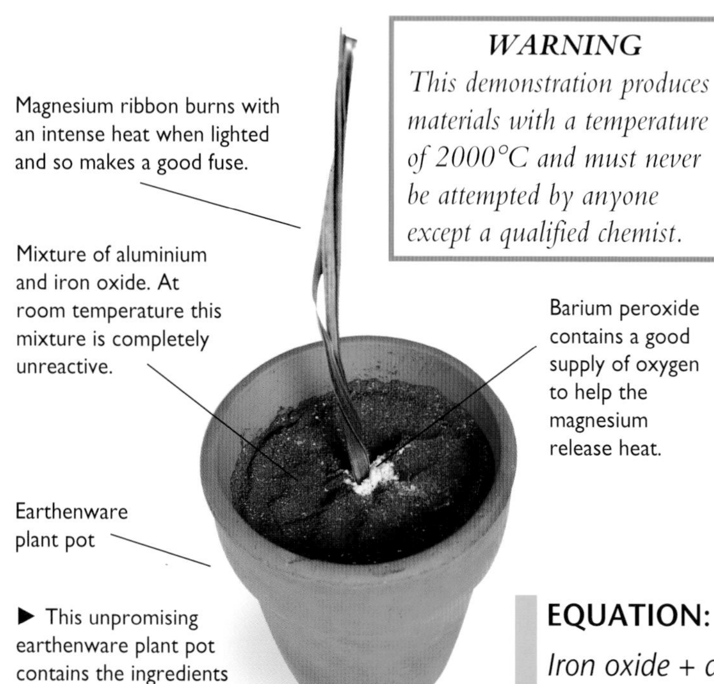

► This unpromising earthenware plant pot contains the ingredients for a spectacular chemical reaction.

Also...

This demonstration shows that materials can be made to react at high temperatures even if they are completely unreactive at low temperatures. A modified form of this demonstration is used to weld rails as they are laid. A bag filled with the powder is tied around the rail sections to be welded and the fuse lit. The rest happens automatically!

The demonstration also shows the principle of extracting iron from its ore. It requires a chemical reaction to strip the oxygen from the iron. However, the use of the equipment shown here is not practical for industrial purposes. Instead, charcoal or coke is used, as shown on pages 24 and 25.

EQUATION: Chemical reduction of iron oxide to iron

Iron oxide + aluminium ⇨ iron metal + aluminium oxide

$$Fe_2O_3(s) + 2Al(s) \Rightarrow 2Fe(s) + Al_2O_3(s)$$

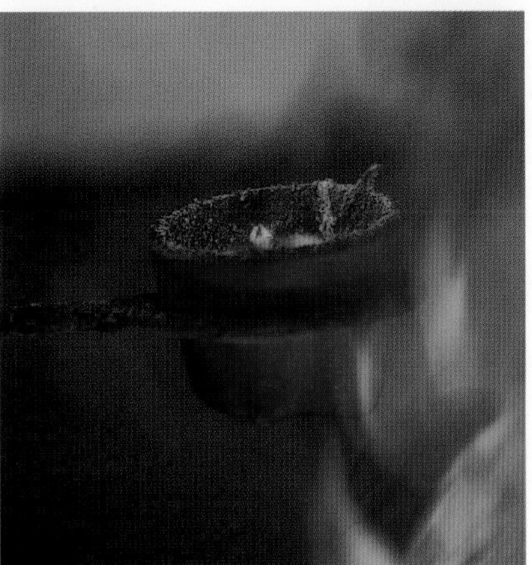

▲▶ The magnesium ribbon heats the barium oxide, starting an intense reaction that releases oxygen and enough heat to ignite the mixture of iron and aluminium. The temperature of the mixture reaches 2000°C and molten iron is released. This falls out of the bottom of the plant pot and is captured in a bath of water. The water in the bath is immediately turned to steam.

decompose: to break down a substance (for example by heat or with the aid of a catalyst) into simpler components. In such a chemical reaction only one substance is involved.

ion: an atom, or group of atoms, that has gained or lost one or more electrons and so developed an electrical charge.

oxidation/reduction: a reaction in which oxygen is gained or lost. (Also… More generally oxidation involves the loss of electrons.)

How it works

❶ The whole apparatus is placed out in a field, well away from any inflammable materials. The fuse is lit. The magnesium burns with a brilliant white light. The fuse burns down into the barium peroxide, causing it to decompose, thus releasing oxygen. Enough heat energy is released to bring the aluminium powder up to the temperature at which it will react.

❷ As soon as the aluminium powder is hot enough, it reacts with oxygen from the iron oxide, releasing more heat. This causes the aluminium to form aluminium oxide, a lightweight, fine powder that is easily carried aloft by the rising currents of heated air. This is what appears as a white smoke.

❸ The reacting aluminium raises the temperature of the iron oxide to about 2000°C. Because the aluminium takes the oxygen from the iron oxide, the iron metal can flow freely.

❹ Iron is heavy, so it sinks through the mixture, flowing out of the hole in the bottom of the plant pot.

❺ The molten iron is collected in a pot filled with water and sand to cool it and prevent it burning its way into the surface below! The water in the container goes up in steam as soon as the iron flows into it, adding to the spectacular effect.

Smelting iron ore

Processing, or smelting, iron ore is done in a blast furnace, a tall oven designed to produce the high temperatures needed to melt and refine iron oxide.

In essence the furnace has to remove the oxygen from the metal ore (the ore must be reduced) and the waste rock must be separated from the valuable metal. This is done with a very hot blast of carbon monoxide gas. The whole furnace is built so that hot carbon monoxide gas is continuously produced and blown through melting rock. The heavy slag and molten iron escape from the bottom of the furnace, and a charge of new ore and fuel is applied at the top. In this way the blast furnace can be operated continuously.

▲ A modern integrated iron and steel works. The raw materials lying on the dockside in front of the furnace are introduced to the top of the furnace by a conveyor system.

► Making iron on a large scale was one of the foundations of the modern industrial world. Giant iron furnaces were built at the heart of many industrial cities.

Over the years, the furnaces became hemmed in by the city, so they moved to sites outside the cities. At the same time they became larger, so that fewer furnaces could make all the iron needed.

This picture shows what an iron furnace looked like in the 19th century.

The modern blast furnace: chemistry in action

Modern blast furnaces are designed to run continuously. They are charged with a mixture of iron ore, coke and limestone at the top of the furnace. Each tonne of iron uses up about three-quarters of a tonne of coke and a quarter of a tonne of limestone.

The blast furnace is designed so that a number of different chemical reactions can occur as the charge moves down through the furnace.

The charge becomes hotter as it moves down, until it melts in the lowest region of the furnace. Iron and waste materials (slag) separate as they melt and are drawn off from the bottom of the furnace.

The iron produced this way is called pig iron. It is the foundation material for other kinds of iron and steel. It is rarely used without further chemical treatment because it is hard and brittle. Cast iron, slightly refined pig iron, can only be used in places where the iron receives little impact or shaking. Cast iron was used as the material for many early iron bridges; steel is used in modern structures.

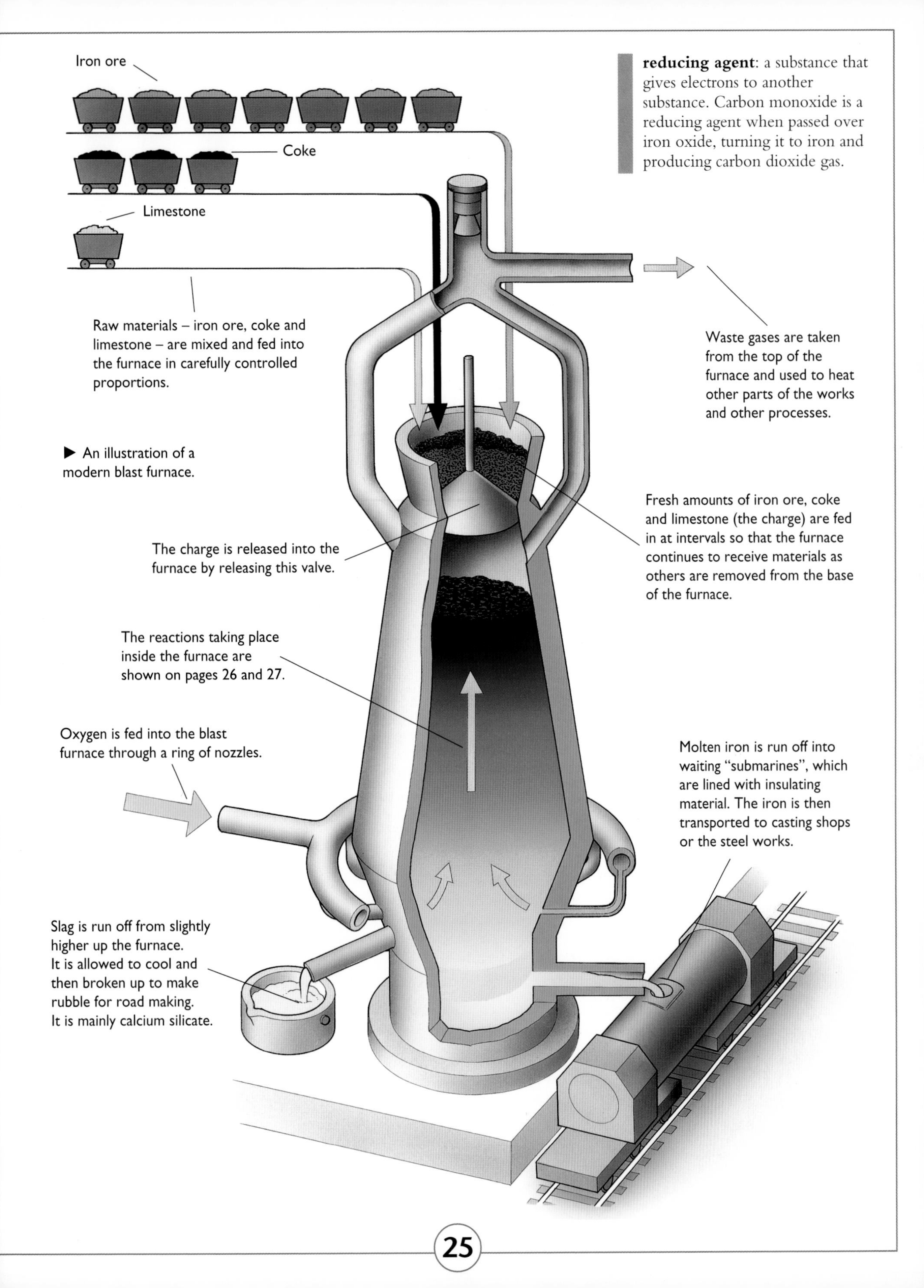

▶ An illustration of a modern blast furnace.

reducing agent: a substance that gives electrons to another substance. Carbon monoxide is a reducing agent when passed over iron oxide, turning it to iron and producing carbon dioxide gas.

The chemistry of smelting

The processes that occur inside the blast furnace are quite complex, and different reactions happen at the top and bottom of the furnace.

The objective is to ensure that the right chemical reactions occur in each part of the furnace so that controlled and continuous processing is achieved. The reactions involved in the general principles of iron-making (shown on the previous page) are described here.

After the last reaction, the iron that flows from the bottom of the furnace into the moulds (known as pigs) is about 95% iron with various proportions of carbon and other impurities.

The waste gases pass out through pipes at the top of the furnace.

Cold iron ore, coke and limestone are added as a charge into the top of the furnace.

Slag

The limestone in the charge decomposes to calcium oxide and gives off carbon dioxide gas. The calcium oxide reacts with the non-metal of the ore. For example, it reacts with the silica of the rock to make calcium silicate.

The mixture of rock materials is known as slag. It is a light grey material, less dense than the iron, which is tapped off the furnace from holes above those used to tap the molten iron.

◀ This picture shows a blast furnace with a pile of slag in front of it. The slag will be allowed to cool and then carried away to be made into useful road-making materials.

The slag, less dense than the iron, is tapped off.

Carbon monoxide: the reducing agent

Carbon monoxide gas is produced at the bottom of the furnace. As air is blown in, the coke, which is almost entirely carbon, is oxidised to produce carbon dioxide gas.

Solid carbon plus oxygen gas from the air react to give carbon dioxide gas, also giving out heat.

As the carbon dioxide bubbles up the furnace, it reacts with more of the coke in the mixture, changing to carbon monoxide gas. This gas then reacts with the iron oxide in the middle of the furnace to give liquid iron.

ore: a rock containing enough of a useful substance to make mining it worthwhile.

reducing agent: a substance that gives electrons to another substance. Carbon monoxide is a reducing agent in a blast furnace.

slag: a mixture of substances that are waste products of a furnace. Most slags are composed mainly of silicates.

Gases pass up through the mixture in the furnace.

Liquids descend through the mixture in the furnace.

Carbon monoxide reacts with iron ore to produce molten iron and carbon dioxide gas. The molten iron sinks to the bottom of the furnace.

The oxygen reacts with the coke to produce carbon dioxide gas, which in turn reacts with more of the coke to produce carbon monoxide.

Oxygen gas is pumped in.

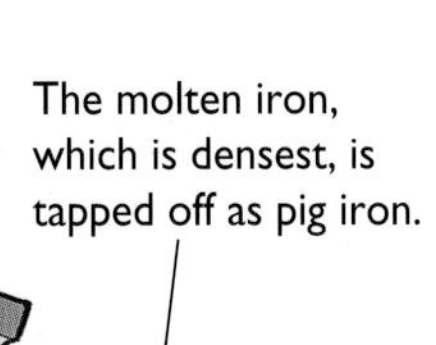

The molten iron, which is densest, is tapped off as pig iron.

EQUATION: Oxidation of coke

carbon + oxygen ⇨ carbon dioxide

$C(s) + O_2(g) \Rightarrow CO_2(g)$

EQUATION: Carbon dioxide reduced

carbon dioxide + carbon ⇨ carbon monoxide

$CO_2(g) + C(s) \Rightarrow 2CO(s)$

Iron ore to iron metal

Iron oxide sinks down the furnace where it reacts with carbon monoxide gas coming up from the bottom of the furnace. The two then react to produce carbon dioxide gas and liquid iron.

EQUATION: Iron oxide is reduced

Iron oxide + carbon monoxide ⇨ iron metal + carbon dioxide

$Fe_2O_3(s) + 3CO(g) \Rightarrow 2Fe(l) + 3CO_2(g)$

◀ When the iron is tapped from the bottom of the furnace, it is often transferred into "submarines". These insulated containers take the molten iron to the nearby steel furnace for conversion to steel. This process is explained on pages 30 and 31.

Cast and wrought iron

The molten iron that comes from a furnace is known as pig iron. This can be run off into moulds to produce a variety of intricate shapes, collectively known as wrought iron. If pig iron is remelted and cooled it forms cast iron. The works in which castings are made is called a foundry.

Iron refined in a furnace is not entirely pure, as it contains carbon. The amount of carbon affects the properties of the iron to a large extent. Wrought iron is nearly pure iron, with a carbon content of less than 0.035%. This makes it a relatively soft material that is easily worked with hammers (it is said to be easily forged).

Cast iron has a carbon content above 3%, which makes it harder than wrought iron but much more brittle. If cast iron is cooled quickly, hard but brittle white cast iron is formed; if it is cooled slowly, soft but tough grey cast iron is formed.

Wrought iron

Wrought iron is nearly pure iron, with a soft, fibrous structure. This allows it to be formed into intricate shapes, whether hot or cold. It was used to make everything from ploughshares to rifles, from railway track to decorative grilles.

Wrought iron replaced bronze in ancient civilisations, and led to the Iron Age. It was originally formed simply by hammering the nearly molten ore to remove the impurities.

All the ironwork in old buildings is wrought iron. Cast iron was only made after the mid-18th century when it replaced wrought iron because it is stronger and can be used more easily. In particular, less material is needed for structural purposes, such as bridges and buildings. Wrought iron continued to be used for railway tracks until it was replaced by steel.

▶ The world's first iron bridge, built across the River Severn in England in 1778, ushered in the new Iron Age. It was made of wrought iron metal. The fastenings are iron wedges rather than rivets because, at this stage, designers were still making bridges of wood and they had not yet learned the technique of riveting.

▲ An 18th-century iron foundry.

Types of cast iron

Cast iron varies according to its silicon and carbon content. Grey cast iron is the most common. It is easily cast and machined, and is used for casting vehicle engine blocks.

To change the properties of grey cast iron into other forms, it must be remelted and the chemical composition altered. Alternatively, some changes can be produced by controlling the rate at which the iron cools after casting.

The uses of cast iron

Cast iron was first introduced as a structural material in the late 18th century. The first buildings to benefit from this new material were factories, where heavy machinery could now be supported on cast iron columns. Warehouses were also quick to use and display this new material.

The idea of supporting a building with cast iron frames was of vital importance to modern society because it led to the development of high rise (skyscraper) frame construction in the 20th century. But some of the most flamboyant uses of the metal were in combination with glass panels to make intricately patterned windows. These were used for railway stations and great exhibition halls such as London's Crystal Palace (built 1851).

For many centuries, ironmakers had made intricate metal objects by working wrought iron with great skill. The mass production of cast iron decorative ware allowed much cheaper production. Eventually the skills of many of these ironworkers were lost.

◀ In the 19th century it became very fashionable to use cast iron as a decorative part of architecture. Many railway stations and other large public buildings show an enormous range of uses for cast iron. However, it was used to great effect in many domestic buildings as well. One of the world's most famous uses of cast iron in buildings is seen on these Sydney houses, where exquisitely produced cast iron decorates many of the townhouses of the period. The cast iron is protected from weathering by paint.

Steelmaking

Steel is mostly made of iron, but many of the impurities have been taken out, and controlled amounts of other compounds have been added.

Iron contains about 5% carbon and small amounts of manganese, silicon, phosphorus and sulphur. The steelmaking process aims to reduce the carbon level to below 1.75% by oxidising iron with oxygen from the air.

Refining the steel

Steel furnaces are the first step in steelmaking. The charge material is collected in a giant steel furnace, where a jet of inert gas such as argon is blown through it so that the steel is stirred and made more uniform. It may then be treated to reduce sulphur and other impurity levels even further. Only then does it go to the mills to be rolled into sheets or made into shapes such as girders, bars and wire.

▼ The basic oxygen furnace process of steelmaking.

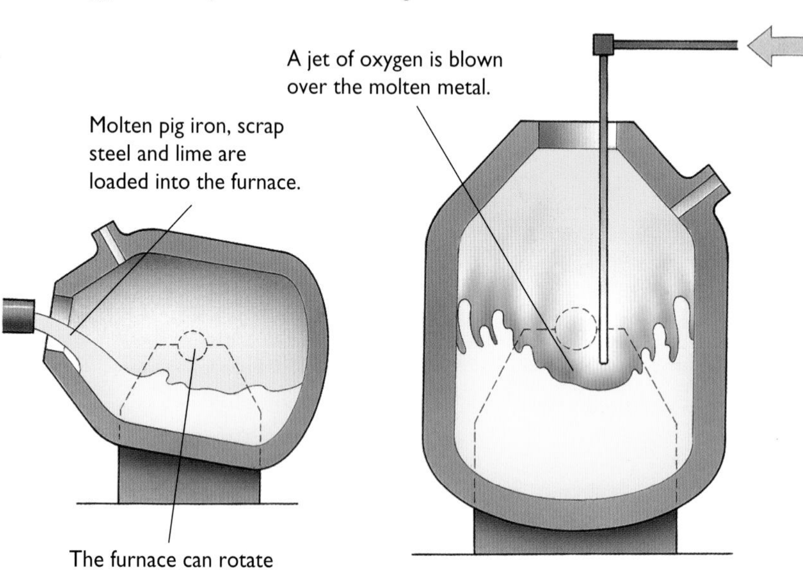

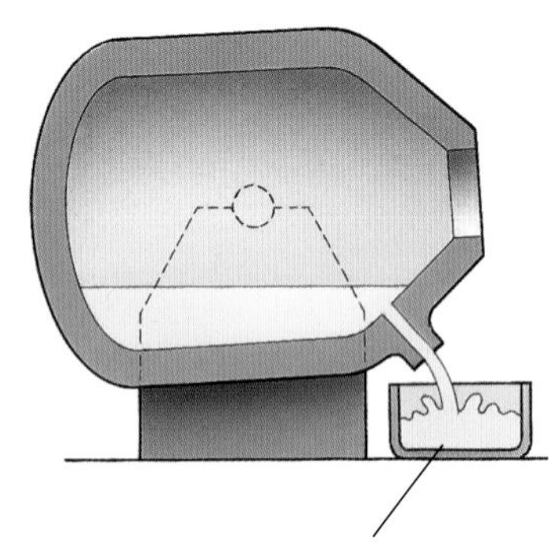

Also...

The electric arc process is another way to produce molten steel, but it is essentially a melting process for scrap steel, and chemical reactions do not play a major part in it.

EQUATION: Iron is oxidised, releasing heat

Iron + oxygen ➪ iron oxide

$2Fe(s) + O_2(g) \Rightarrow 2FeO(s)$

heat is released

EQUATION: Silicon impurities are oxidised using lime

Iron oxide + silicon + lime ➪ steel + slag

$2FeO(s) + Si(s) + CaO(s) \Rightarrow 2Fe(s) + CaSiO_3(s)$

EQUATION: Excess carbon in the iron is removed

Iron oxide + carbon in the iron ➪ steel + carbon monoxide gas

$FeO(s) + C(s) \Rightarrow Fe(s) + CO(g)$

The steel furnace

Just as the iron-making process is chemistry on a grand scale, so too is the steelmaking process.

The steel furnace is where the charge of scrap steel and iron is heated, and where the chemical reactions occur. The first steel furnace was the Bessemer converter, invented by Henry Bessemer in the 1850s. It uses a large pot lined with a basic material such as limestone (calcium carbonate). A blast of air is introduced into the molten iron, causing the iron to oxidise and release a large amount of heat.

As the temperature in the steel furnace rises, the limestone decomposes, releasing calcium oxide. The impurities in the iron react with the calcium oxides to form a molten slag. At this stage the furnace "blows": huge flames appear at the furnace mouth and gases boil up through the liquid metal.

The oxygen from the iron combines with excess carbon in the iron, oxidising it to carbon monoxide. In this way the high carbon content of the iron is lowered, and impurities left in the iron from the blast furnace stage are removed.

The Bessemer converter has been improved and is now called the basic oxygen furnace process. However, the chemical reactions that take place remain the same.

▲ Inside a steel-rolling mill.

oxidation: a reaction in which the oxidising agent removes electrons.

▲ Sir Henry Bessemer.

▲ This engraving shows a Bessemer converter in use in the last century.

▼ An integrated iron and steel works, such as this one in Port Kembla, New South Wales, can be regarded as a giant chemical works.

Raw materials on the dock side

Blast furnace

Steel furnace

Mild steel

Mild steel – steel with a carbon content of about 0.25% – is one of the most widely used materials today. It is strong and readily available. Its main drawback is that it is a reactive metal and so corrodes readily. This means that all steel objects that are to be used where moisture is present must be protected in some way. Mild steel is also naturally quite easily bent, which is an advantage for making shapes, but a problem when using steel in vehicle panels. To overcome this problem the steel needs to be hardened.

▼ Despite being constantly exposed to moisture, all ships, and most containers, are still made from steel. This is because the cost of using other materials would be vastly greater. As with most uses of steel, manufacturers have to compromise between resistance to corrosion and cost. As a result, much research goes into protective coatings. Many steel structures are also protected electrochemically by cathodic protection.

▼ Steel is widely used in the construction industry, for example in reinforced concrete buildings. Here the strength of the steel helps to support the concrete. The steel mesh and rods used do not corrode, because once the concrete has set around them the steel is protected from the air and this prevents chemical corrosion.

▼ Steel will take a cutting edge well and is the main material used for making chisels, saw blades, twist drills and many other tools. Because it is not as hard as some other forms of steel, cutting tools require frequent sharpening.

▶ One of the major steel uses in the 19th century was in railway lines: one-quarter of a million miles of track were laid at this time.

▼ Steel is used as the body panels of almost all vehicles. It is a strong material, especially when formed into complex shapes, and it is relatively inexpensive. It can be formed into shapes when cold.

cathodic protection: the technique of making the object that is to be protected from corrosion into the cathode of a cell. For example, a material, such as steel, is protected by coupling it with a more reactive metal, such as magnesium. Steel forms the cathode and magnesium the anode. Zinc protects steel in the same way.

Hardening mild steel

Mild steel is relatively soft and for many purposes it must be hardened. In steel the hardness depends on the amount of carbon present. Carbon becomes part of the structure of steel, making an alloy of iron and carbon atoms. The structure of the atoms can be layered, and the properties changed by heating the steel in molten sodium cyanide. This makes the steel absorb nitrogen and carbon atoms which, when the steel is cooled quickly, causes the surface to harden greatly. This is known as case hardening.

Steel can also be hardened by tempering. This is the process of reheating the steel and then cooling it without bringing about any chemical change. This process also makes the steel tougher and less liable to crack.

Special steels

Steel is an extremely versatile material, whose properties can be tailored to a wide range of needs. For example, by changing the amount of carbon in the steel, the strength can be improved. In general the higher the carbon content, the tougher the steel. Thus, ordinary (mild) steel (described on page 32) contains 0.25% carbon; medium steels contain 0.5% carbon and high carbon steels (the toughest) contain from 0.6% to 2% carbon. Varying the chemical content in this way is quite different to heating the metal and then cooling it quickly (quenching), a process that hardens the steel (see the previous page).

The reason that not all steels have a high carbon content is that it makes the steel more brittle. This is why mild steel is often preferred for example, for pressing vehicle panels. But medium steels are best for beams and other construction uses and high carbon steels are preferred for machine tools.

Adding other metals to a steel also causes it to become harder. Manganese and silicon are two of the most commonly used alloying metals. Tungsten is an important alloying metal when hardness needs to be maintained even at very high temperatures, such as at the tip of a high-speed drill cutting into metal. Chromium and nickel as alloying elements are used to improve corrosion resistance, giving stainless steels. Silicon steel can be magnetised. Cobalt has a similar effect, and is used to make permanent magnets. Finally, molybdenum and vanadium are also used as alloying metals for special steels.

alloy: a mixture of a metal and various other elements.

◀ Gateway Arch, St. Louis, Missouri, USA, designed by Eero Saarinen, is a monument to cladding made from stainless steel. It is 192 metres high and took nearly two years to build.

Stainless steel

Stainless steels are a range of alloys based on iron. The name comes from the property of such steel to resist reactions with many other substances which may cause surface staining and rusting.

The most common stainless steels are iron-based alloys with a very high resistance to rusting and corrosion. This is because of their chromium content, which is greater than that found in other types of steel.

The most common stainless steel is made with 18% chromium, 8% nickel and 0.15% carbon. This is the type of stainless steel used for making everyday items such as cutlery.

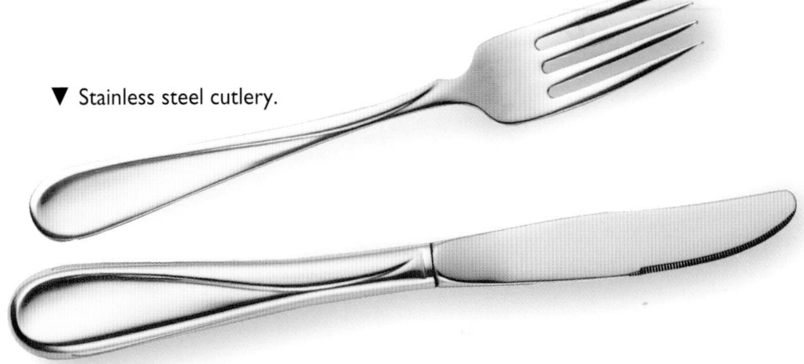

▼ Stainless steel cutlery.

▲ Saws need to remain sharp for as long as possible. Many of the best-wearing blades are made from tungsten steel. The tungsten makes the steel considerably harder and it is used in, for example, hard-wearing drills.

Chromium

Chromium was named after "chroma", the Greek word for colour, because of the many coloured compounds that it makes. Among these are the rich red of rubies and the brilliant green of emeralds.

The main ore of chromium is chromite, found in combination with some iron ores.

Chromium ore is refined into shiny, silvery-white chromium metal. However, because chromium metal is brittle and hard, it has no uses on its own.

Chrome alloys

Chromium is used as an alloy with steel. Chrome steels are strong and hard and are often used for making tools. Chromium is always a component of stainless steel.

Alloys of nickel and chromium are used for high temperature wire elements in electric fires and other appliances.

▼ A piece of highly reflective chromium metal.

Chromium plating

Because chromium resists corrosion, is hard-wearing and has an attractive silvery colour that does not tarnish, it is widely used to plate iron and brass.

All items to be chromium plated are dipped in a bath containing chromium compounds in solution. In some cases, the articles to be plated are simply dipped in chromium metal. But a thin, even finish is created by using the process of electrolysis. In this case the articles to be plated are attached to the negative end of an electrical circuit, and the surface is then plated.

Corrosion inhibitors

Because chromium compounds resist corrosion, they can be used as corrosion inhibitors in water cooling systems, such as that of a motor vehicle.

alloy: a mixture of a metal and various other elements.

electrolysis: an electrical–chemical process that uses an electric current to cause the break up of a compound and the movement of metal ions in a solution. The process happens in many natural situations (as for example in rusting) and is also commonly used in industry for purifying (refining) metals or for plating metal objects with a fine, even metal coating.

electroplating: depositing a thin layer of a metal onto the surface of another substance using electrolysis.

◀ Chromium was used as a decorative finish in many of the world's older classic cars. This is because, at the time, no other corrosion-resistant finish was available. Modern bumpers are not made of iron, but of polycarbonate, and so do not need to be chromium protected at all.

Preparation of chromic acid

Chromic acid is prepared by stirring brown chromic oxide in water. The resulting acid is very corrosive and is used for cleaning glassware. It is also a strong oxidising agent.

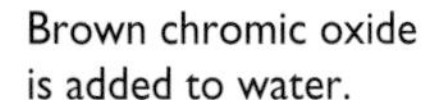

EQUATION: Laboratory production of chromic acid

Chromic oxide + water ➪ chromic acid

$CrO_3(s)$ + $H_2O(l)$ ➪ $H_2CrO_4(aq)$

Chromium colours

Chromium compounds are mainly brightly coloured. As a result, they are widely used as the colouring agent, or pigment, in paints, glazes, glass and enamels.

The most common colours for chromium compounds are green, yellow and orange.

Glaze

An important use of coloured chrome oxide is to colour glaze. A glaze is a mixture of colouring powder and powdered glass in water. The article to be glazed is then dipped in the mixture, or it may be sprayed or brushed on.

The article is then fired so that the glass melts and forms a surface glaze.

Also...

Paint manufacturers are among the main users of chromium compounds. A paint must protect the surface to which it is applied and also provide a pleasing and decorative colour. It must be easy to apply, make a thin even coating, and then dry quickly and uniformly.

Every paint is made of two parts, a powder of solid particles that gives the colour and also makes the paint opaque. This is known as the pigment. The pigment is mixed into a liquid, called the binder, that allows the pigment to be spread evenly over the surface.

In some cases the binder is made from a solvent such as those based on petroleum. When applied, this evaporates and leaves a deposit of the pigment on the surface. At the same time the binder reacts with oxygen of the air, becoming a non-sticky solid.

In many more cases, modern paints are being made with a water base, for environmental reasons. However, pigments like chromium compounds, do not easily mix with water and must be added already wrapped in their binder. This suspension of particles is called an emulsion.

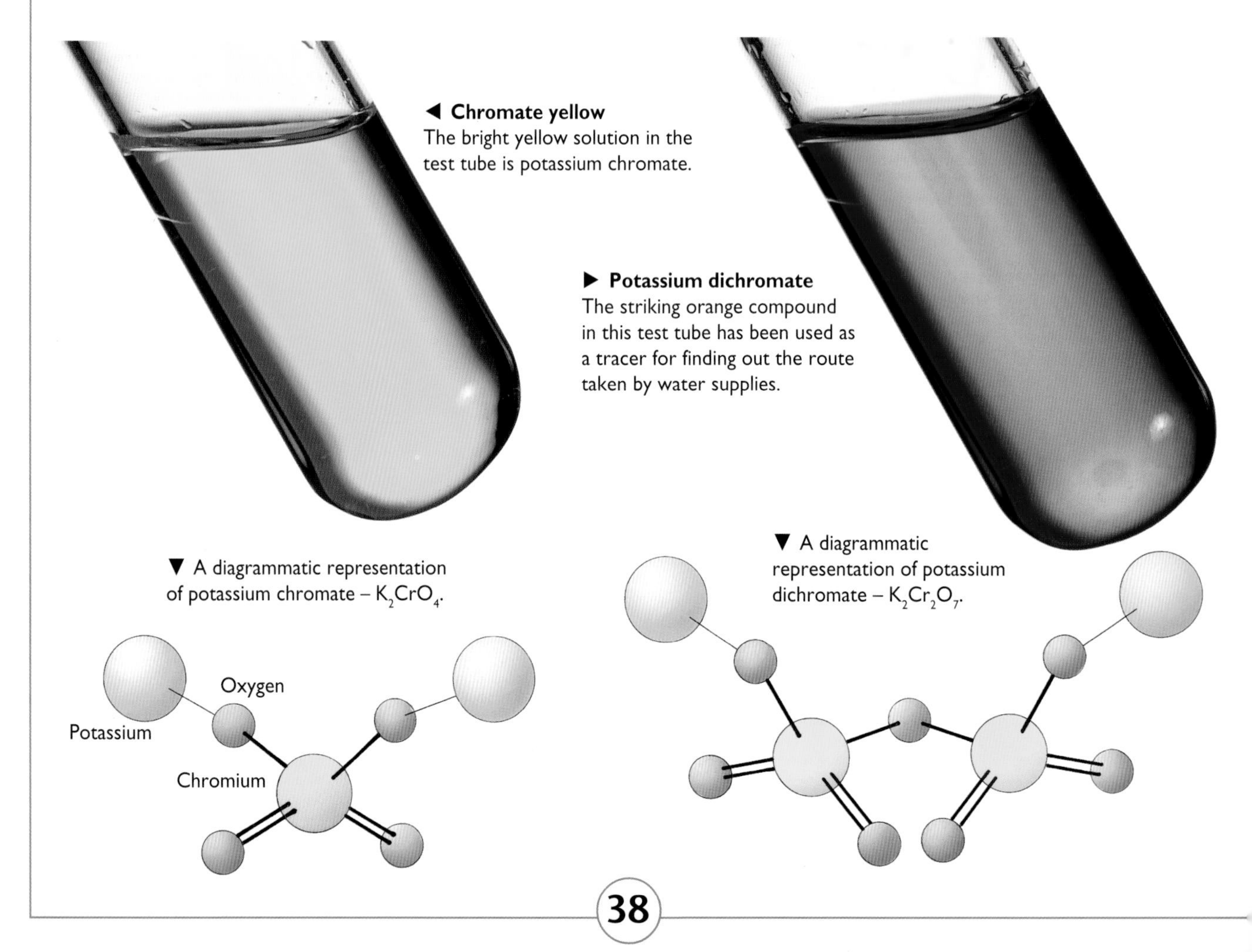

◀ **Chromate yellow**
The bright yellow solution in the test tube is potassium chromate.

▶ **Potassium dichromate**
The striking orange compound in this test tube has been used as a tracer for finding out the route taken by water supplies.

▼ A diagrammatic representation of potassium chromate – K_2CrO_4.

▼ A diagrammatic representation of potassium dichromate – $K_2Cr_2O_7$.

▶ **Chromium III oxide**
This green substance is used as a pigment for paint.

▼ This is a yellow precipitate of lead chromate. It was prepared by adding yellow potassium chromate solution to colourless lead nitrate solution. The beaker is sitting on a magnetic stirrer that spins the plastic coated metal mixer (sometimes called a "magnetic flea") in the bottom. This thoroughly stirs the chemicals.
In this case the potassium chromate solution was added about a second before the picture was taken.

EQUATION: Reaction of potassium chromate and lead nitrate

Potassium chromate + lead II nitrate ⇨ lead chromate + potassium nitrate

$K_2CrO_4(aq) + Pb(NO_3)_2(aq) \Rightarrow PbCrO_4(s) + 2KNO_3(aq)$

Manganese

Manganese is a grey metal that looks very much like iron. Pure manganese is rarely used because it is hard and brittle. Its main use is as an alloying and purifying metal in the iron and steel industry.

Manganese is more reactive than iron (it is higher up the reactivity series), so when it is added to a furnace, the iron oxides react with the manganese, removing the oxygen from the iron and adding it to the manganese. Manganese oxide is one of the constituents of slag.

Manganese also reacts with sulphur impurities to form manganese sulphide, which then adds to the slag.

Manganese can be added to purified iron to make an alloy. The alloy is tougher, yet easier to work into shape, than pure steel. It also gives added corrosion resistance. Manganese steel contains just over one-tenth manganese.

The alloy has the added advantage of being nonmagnetic. Its toughness means that it is used for the buckets and scoops of heavy lifting machinery and many other uses within the construction industry.

When manganese makes the main component of an alloy with nonferrous metals like copper, it produces metals that expand and contract readily with temperature. As a result they are used in bimetal strips as heat-sensitive switches.

▲ A Venetian glass vase coloured using compounds of manganese.

Manganese oxide in glass-making

The Venetians were the first people to use manganese compounds in glass-making. In the Middle Ages they discovered that manganese oxide would oxidise the iron impurities in the glass that gave it a green or brown tint. The result was a much clearer, and therefore more popular, glass.

▼ Garnet is a silicate in which manganese and iron are often important constituents, producing dark red–brown crystals. Garnet is widely regarded as a semiprecious stone.

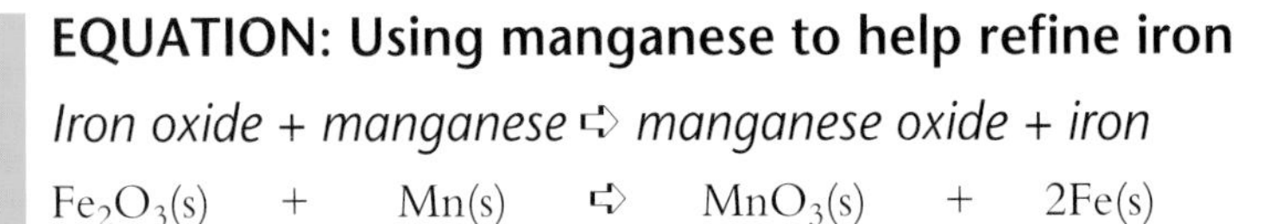

EQUATION: Using manganese to help refine iron

Iron oxide + manganese ⇨ manganese oxide + iron

$Fe_2O_3(s)$ + $Mn(s)$ ⇨ $MnO_3(s)$ + $2Fe(s)$

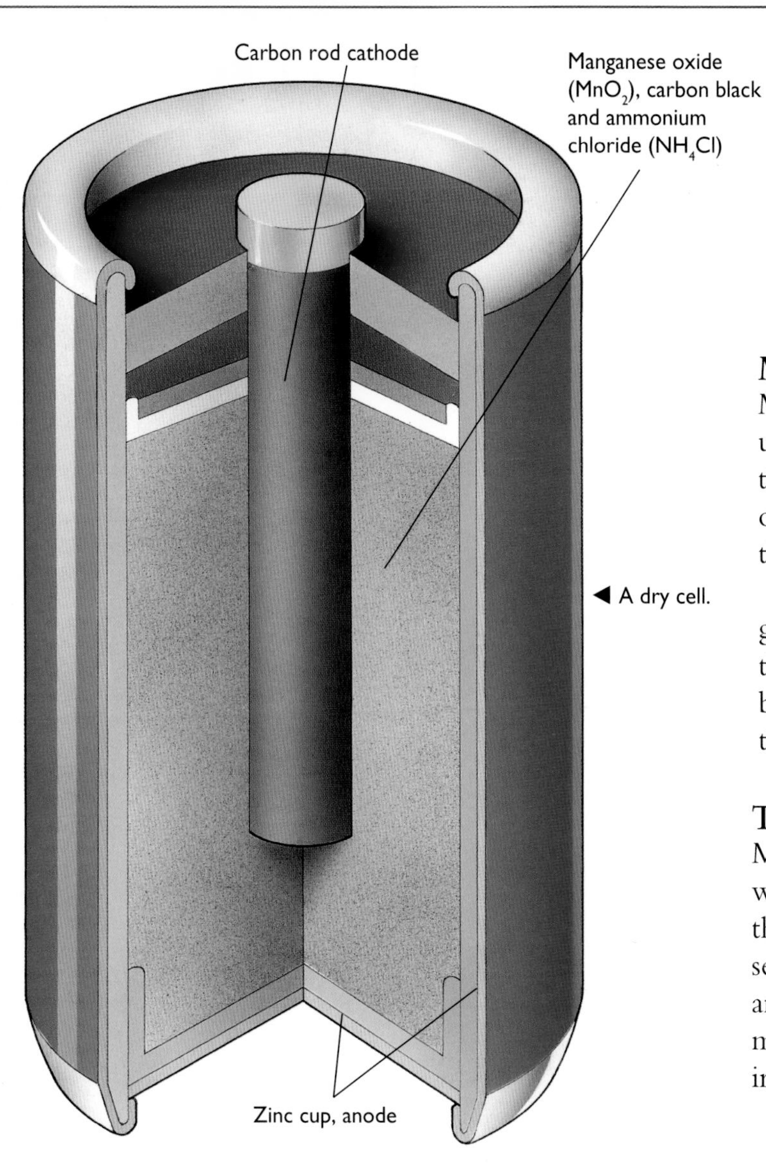

◀ A dry cell.

Manganese nodules

Manganese nodules are some of the richest concentrations of any element. The nodules contain mainly manganese and iron oxides, with some other metals such as cobalt and nickel.

The manganese nodules form naturally from metal compounds contained in materials sinking down through sea waters. Near the sea floor the manganese precipitates out of the sea water as small grains. These grains slowly attract more manganese compounds and eventually grow into nodules. At least half the mass of a manganese nodule is made of manganese.

Manganese nodules are found exclusively on the ocean floor, often lying scattered in huge fields. Each nodule can be the size of a golf ball.

corrosion: the *slow* decay of a substance resulting from contact with gases and liquids in the environment. The term is often applied to metals. Rust is the corrosion of iron.

precipitate: tiny solid particles formed as a result of a chemical reaction between two liquids or gases.

Manganese oxide in batteries

Manganese oxide is the most commonly used compound of manganese. Under the name of pyrolusite, manganese oxide (a black powder) is bound in with the carbon electrodes of a dry battery.

Manganese oxide "traps" hydrogen gas that is given off as the chemicals of the battery react. In this way the dry battery does not swell up and burst from the formation of hydrogen gas.

The reactivity of manganese

Metals can be arranged in a reactivity series, with the most reactive at the top and the least reactive at the bottom. As you can see, manganese comes above iron in the list and so is more reactive. This means that manganese will remove the oxygen from iron in a heated furnace.

REACTIVITY SERIES	
Element	*Reactivity*
potassium	*most reactive*
sodium	
calcium	
magnesium	
aluminium	
manganese	
chromium	
zinc	
iron	
cadmium	
tin	
lead	
copper	
mercury	
silver	
gold	
platinum	*least reactive*

Potassium permanganate

Manganese has compounds with striking colours. Some are used in the paint industry as pigments. Manganese sulphate is bright red and is used to colour enamels; manganese carbonate is white.

When a manganese compound reacts with potassium nitrate, a deep green compound (potassium manganate) results. Further reaction with sulphuric acid produces purple potassium permanganate (an oxidising agent used for bleaching and as a mild disinfectant).

◀ One of the most widely used compounds of manganese is potassium permanganate. It is a mild disinfectant. It is water soluble and has a strong colour, making it suitable for use as a tracing dye.

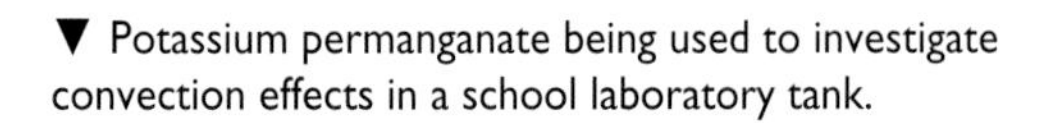

▼ Potassium permanganate being used to investigate convection effects in a school laboratory tank.

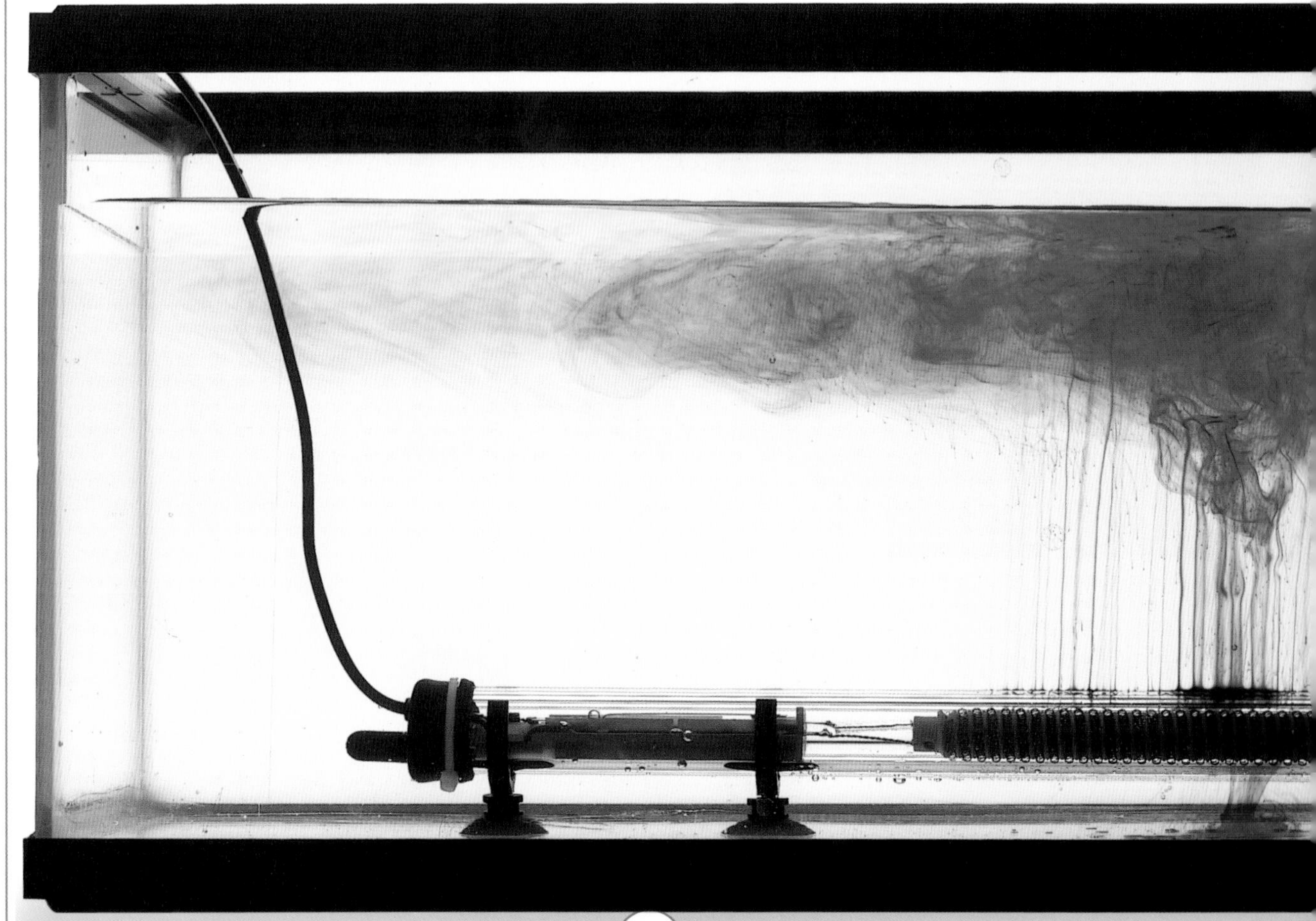

▼ Manganese oxide, potassium nitrate and potassium hydroxide at the start of the demonstration.

oxidising agent: a substance that removes electrons from another substance (and therefore is itself reduced).

▼ Potassium manganate, a deep green compound, is produced after heating and stirring.

The preparation of potassium permanganate

Black manganese oxide powder is mixed with potassium nitrate and pellets of solid potassium hydroxide. Heated and stirred, they produce potassium manganate. When acidified and dried, this produces potassium permanganate.

Key facts about...

Iron

A soft, silver–grey metal, chemical symbol Fe

Essential for plant growth

Fourth most plentiful element at the surface of the Earth

Rusts easily

Has no smell

Forms the colouring of most rocks

Has no taste

Magnetic

Makes up over 5% of the Earth's crust

Easily forms alloys with other metals

Found most commonly as iron oxide in rocks

Atomic number 26, atomic weight about 56

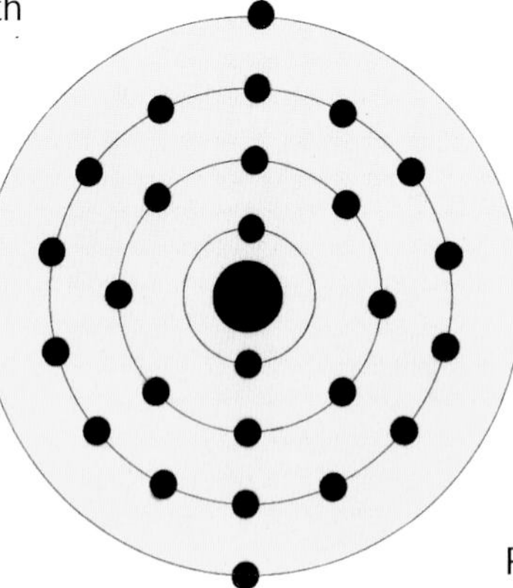

SHELL DIAGRAMS

The shell diagrams on these two pages are representations of an atom of each element. The total number of electrons is shown in the relevant orbitals, or shells, around the central nucleus.

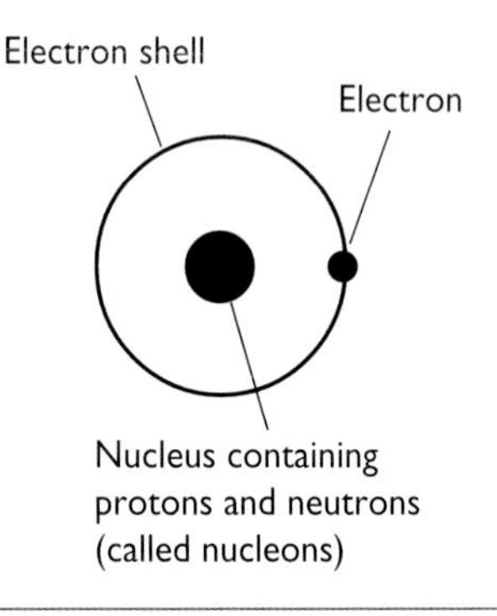

▶ The 19th-century wonders of the engineering world were based on iron.

Chromium

A soft, silver–white metal, chemical symbol Cr

Rare metal that does not occur uncombined in nature

Many of the compounds are toxic

Highly reflective

Has no smell

Has no taste

Atomic number 24, atomic weight about 52

Alloyed with other metals to improve their strength and corrosion resistance

Compounds are brightly coloured

Resistant to corrosion

Essential trace element in food

Manganese

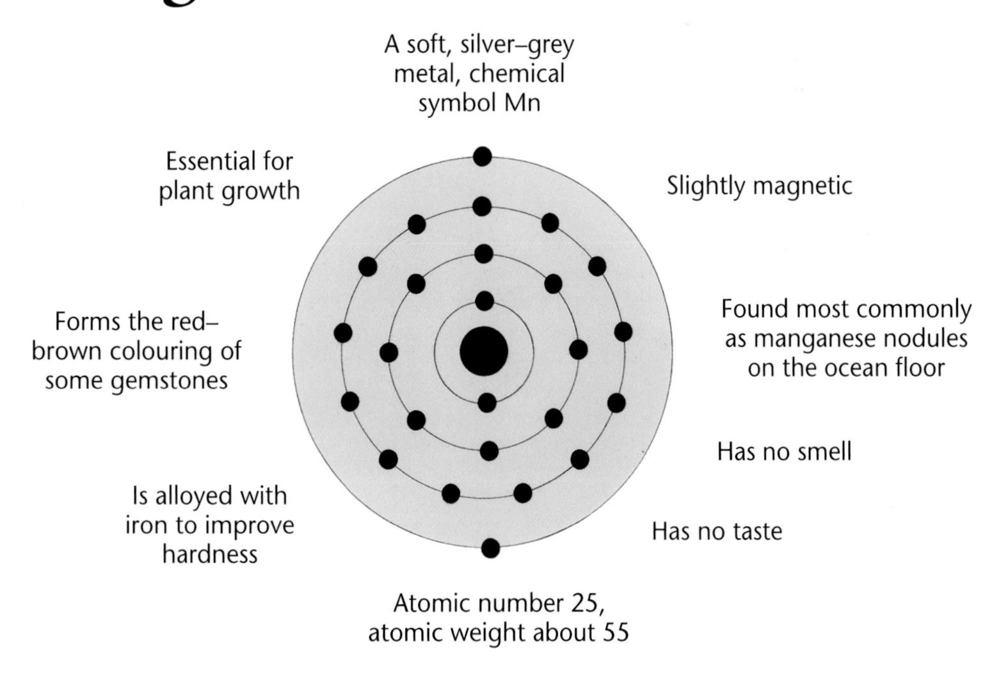

The Periodic Table

The Periodic Table sets out the relationships among the elements of the Universe. According to the Periodic Table, certain elements fall into groups. The pattern of these groups has, in the past, allowed scientists to predict elements that had not at that time been discovered. It can still be used today to predict the properties of unfamiliar elements.

The Periodic Table was first described by a Russian teacher, Dmitry Ivanovich Mendeleev, between 1869 and 1870. He was interested in writing a chemistry textbook, and wanted to show his students that there were certain patterns in the elements that had been discovered. So he set out the elements (of which there were 57 at the time) according to their known properties. On the assumption that there was pattern to the elements, he left blank spaces where elements seemed to be missing. Using this first version of the Periodic Table, he was able to predict in detail the chemical and physical properties of elements that had not yet been discovered. Other scientists began to look for the missing elements, and they soon found them.

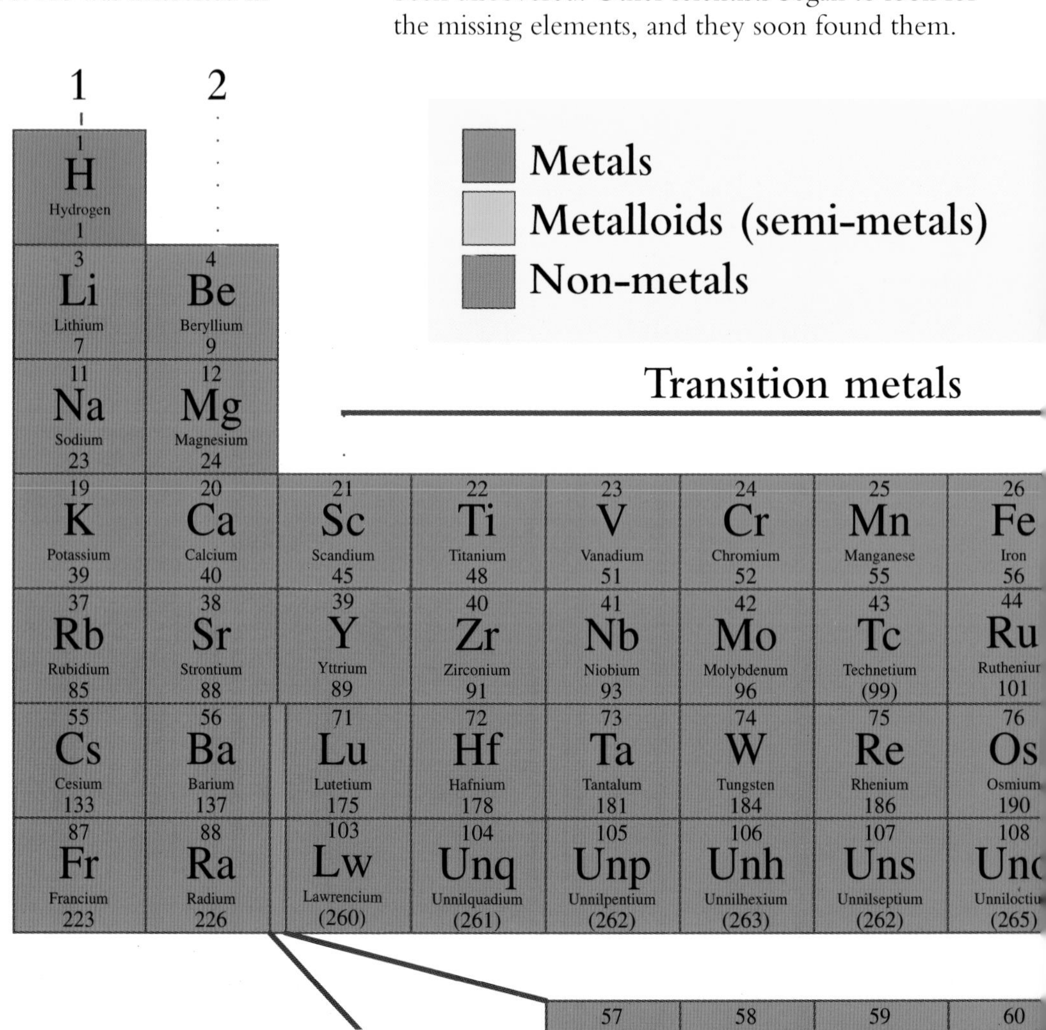

Hydrogen did not seem to fit into the table, so he placed it in a box on its own. Otherwise the elements were all placed horizontally. When an element was reached with properties similar to the first one in the top row, a second row was started. By following this rule, similarities among the elements can be found by reading up and down. By reading across the rows, the elements progressively increase their atomic number. This number indicates the number of positively charged particles (protons) in the nucleus of each atom. This is also the number of negatively charged particles (electrons) in the atom.

The chemical properties of an element depend on the number of electrons in the outermost shell.

Atoms can form compounds by sharing electrons in their outermost shells. This explains why atoms with a full set of electrons (like helium, an inert gas) are unreactive, whereas atoms with an incomplete electron shell (such as chlorine) are very reactive. Elements can also combine by the complete transfer of electrons from metals to non-metals and the compounds formed contain ions.

Radioactive elements lose particles from their nucleus and electrons from their surrounding shells. As a result their atomic number changes and they become new elements.

Atomic (proton) number: 13
Symbol: Al
Name: Aluminium
Approximate relative atomic mass (Approximate atomic weight): 27

				3	4	5	6	7	0
									2 He Helium 4
				5 B Boron 11	6 C Carbon 12	7 N Nitrogen 14	8 O Oxygen 16	9 F Fluorine 19	10 Ne Neon 20
				13 Al Aluminium 27	14 Si Silicon 28	15 P Phosphorus 31	16 S Sulphur 32	17 Cl Chlorine 35	18 Ar Argon 40
27 Co Cobalt 59	28 Ni Nickel 59	29 Cu Copper 64	30 Zn Zinc 65	31 Ga Gallium 70	32 Ge Germanium 73	33 As Arsenic 75	34 Se Selenium 79	35 Br Bromine 80	36 Kr Krypton 84
45 Rh hodium 103	46 Pd Palladium 106	47 Ag Silver 108	48 Cd Cadmium 112	49 In Indium 115	50 Sn Tin 119	51 Sb Antimony 122	52 Te Tellurium 128	53 I Iodine 127	54 Xe Xenon 131
77 Ir Iridium 192	78 Pt Platinum 195	79 Au Gold 197	80 Hg Mercury 201	81 Tl Thallium 204	82 Pb Lead 207	83 Bi Bismuth 209	84 Po Polonium (209)	85 At Astatine (210)	86 Rn Radon (222)
109 Jne nilennium (266)									

61 Pm omethium (145)	62 Sm Samarium 150	63 Eu Europium 152	64 Gd Gadolinium 157	65 Tb Terbium 159	66 Dy Dysprosium 163	67 Ho Holmium 165	68 Er Erbium 167	69 Tm Thulium 169	70 Yb Ytterbium 173
93 Np eptunium (237)	94 Pu Plutonium (244)	95 Am Americium (243)	96 Cm Curium (247)	97 Bk Berkelium (247)	98 Cf Californium (251)	99 Es Einsteinium (252)	100 Fm Fermium (257)	101 Md Mendelevium (258)	102 No Nobelium (259)

Understanding equations

As you read through this book, you will notice that many pages contain equations using symbols. If you are not familiar with these symbols, read this page. Symbols make it easy for chemists to write out the reactions that are occurring in a way that allows a better understanding of the processes involved.

Symbols for the elements

The basis of the modern use of symbols for elements dates back to the 19th century. At this time a shorthand was developed using the first letter of the element wherever possible. Thus "O" stands for oxygen, "H" stands for hydrogen and so on. However, if we were to use only the first letter, then there could be some confusion. For example, nitrogen and nickel would both use the symbols N. To overcome this problem, many elements are symbolised using the first two letters of their full name, and the second letter is lowercase. Thus although nitrogen is N, nickel becomes Ni. Not all symbols come from the English name; many use the Latin name instead. This is why, for example, gold is not G but Au (for the Latin *aurum*) and sodium has the symbol Na, from the Latin *natrium*.

Compounds of elements are made by combining letters. Thus the molecule carbon

Written and symbolic equations

In this book, important chemical equations are briefly stated in words (these are called word equations), and are then shown in their symbolic form along with the states.

What reaction the equation illustrates

EQUATION: The formation of calcium hydroxide

Word equation — *Calcium oxide + water ⇨ calcium hydroxide*

Symbol equation — $CaO(s) + H_2O(l) \Rightarrow Ca(OH)_2(aq)$
heated

Sometimes you will find additional descriptions below the symbolic equation.

Symbol showing the state: *s* is for solid, *l* is for liquid, *g* is for gas and *aq* is for aqueous.

Diagrams

Some of the equations are shown as graphic representations.

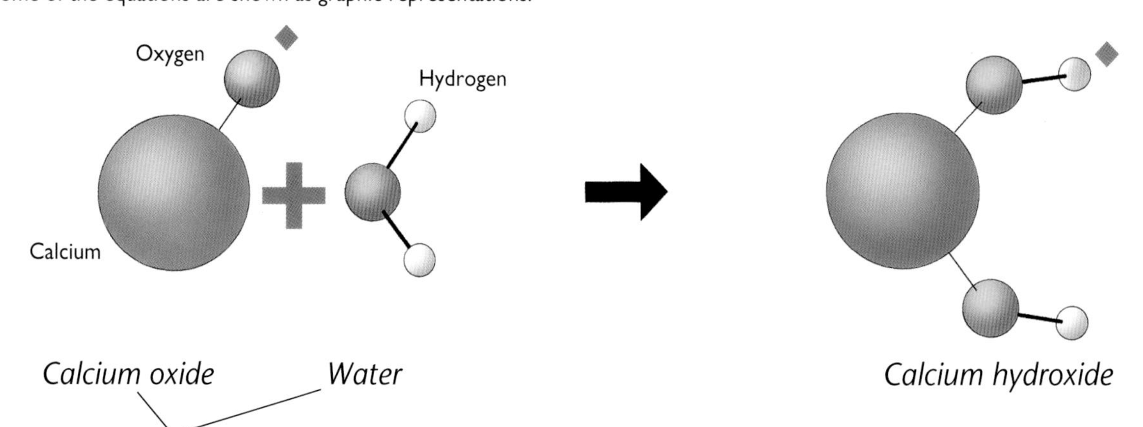

Sometimes the written equation is broken up and put below the relevant stages in the graphic representation.

monoxide is CO. By using lowercase letters for the second letter of an element, it is possible to show that cobalt, symbol Co, is not the same as the molecule carbon monoxide, CO.

However, the letters can be made to do much more than this. In many molecules, atoms combine in unequal numbers. So, for example, carbon dioxide has one atom of carbon for every two of oxygen. This is shown by using the number 2 beside the oxygen, and the symbol becomes CO_2.

In practice, some groups of atoms combine as a unit with other substances. Thus, for example, calcium bicarbonate (one of the compounds used in some antacid pills) is written $Ca(HCO_3)_2$. This shows that the part of the substance inside the brackets reacts as a unit and the "2" outside the brackets shows the presence of two such units.

Some substances attract water molecules to themselves. To show this a dot is used. Thus the blue form of copper sulphate is written $CuSO_4.5H_2O$. In this case five molecules of water attract to one of copper sulphate.

When you see the dot, you know that this water can be driven off by heating; it is part of the crystal structure.

In a reaction substances change by rearranging the combinations of atoms. The way they change is shown by using the chemical symbols, placing those that will react (the starting materials, or reactants) on the left and the products of the reaction on the right. Between the two, chemists use an arrow to show which way the reaction is occurring.

It is possible to describe a reaction in words. This gives word equations, which are given throughout this book. However, it is easier to understand what is happening by using an equation containing symbols. These are also given in many places. They are not given when the equations are very complex.

In any equation both sides balance; that is, there must be an equal number of like atoms on both sides of the arrow. When you try to write down reactions, you, too, must balance your equation; you cannot have a few atoms left over at the end!

The symbols in brackets are abbreviations for the physical state of each substance taking part, so that (*s*) is used for solid, (*l*) for liquid, (*g*) for gas and (*aq*) for an aqueous solution, that is, a solution of a substance dissolved in water.

Atoms and ions
Each sphere represents a particle of an element. A particle can be an atom or an ion. Each atom or ion is associated with other atoms or ions through bonds – forces of attraction. The size of the particles and the nature of the bonds can be extremely important in determining the nature of the reaction or the properties of the compound.

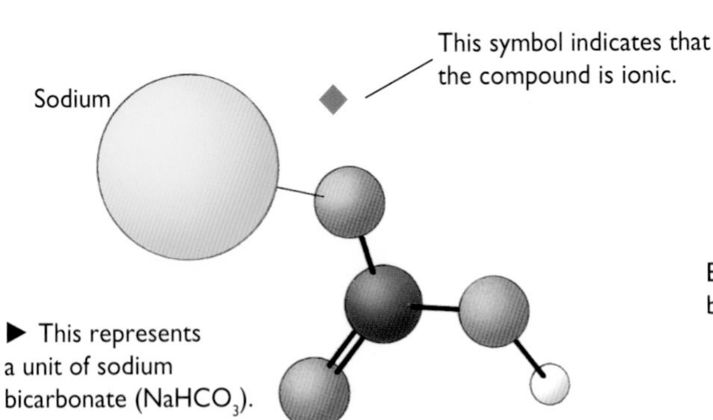

▶ This represents a unit of sodium bicarbonate ($NaHCO_3$).

The term "unit" is sometimes used to simplify the representation of a combination of ions.

Chemical symbols, equations and diagrams
The arrangement of any molecule or compound can be shown in one of the two ways shown below, depending on which gives the clearer picture. The left-hand diagram is called a ball-and-stick diagram because it uses rods and spheres to show the structure of the material. This example shows water, H_2O. There are two hydrogen atoms and one oxygen atom.

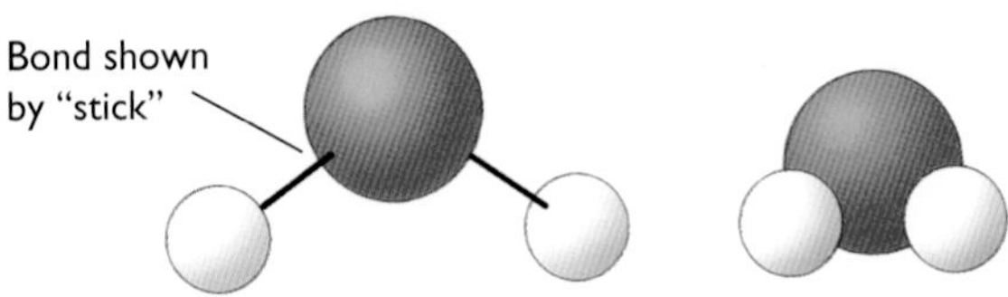

Colours too
The colours of each of the particles help differentiate the elements involved. The diagram can then be matched to the written and symbolic equation given with the diagram. In the case above, oxygen is red and hydrogen is grey.

Glossary of technical terms

absorb: to soak up a substance. Compare to adsorb.

acetone: a petroleum-based solvent.

acid: compounds containing hydrogen which can attack and dissolve many substances. Acids are described as weak or strong, dilute or concentrated, mineral or organic.

acidity: a general term for the strength of an acid in a solution.

acid rain: rain that is contaminated by acid gases such as sulphur dioxide and nitrogen oxides released by pollution.

adsorb/adsorption: to "collect" gas molecules or other particles on to the *surface* of a substance. They are not chemically combined and can be removed. (The process is called "adsorption".) Compare to absorb.

alchemy: the traditional "art" of working with chemicals that prevailed through the Middle Ages. One of the main challenges of alchemy was to make gold from lead. Alchemy faded away as scientific chemistry was developed in the 17th century.

alkali: a base in solution.

alkaline: the opposite of acidic. Alkalis are bases that dissolve, and alkaline materials are called basic materials. Solutions of alkalis have a pH greater than 7.0 because they contain relatively few hydrogen ions.

alloy: a mixture of a metal and various other elements.

alpha particle: a stable combination of two protons and two neutrons, which is ejected from the nucleus of a radioactive atom as it decays. An alpha particle is also the nucleus of the atom of helium. If it captures two electrons it can become a neutral helium atom.

amalgam: a liquid alloy of mercury with another metal.

amino acid: amino acids are organic compounds that are the building blocks for the proteins in the body.

amorphous: a solid in which the atoms are not arranged regularly (i.e. "glassy"). Compare with crystalline.

amphoteric: a metal that will react with both acids and alkalis.

anhydrous: a substance from which water has been removed by heating. Many hydrated salts are crystalline. When they are heated and the water is driven off, the material changes to an anhydrous powder.

anion: a negatively charged atom or group of atoms.

anode: the negative terminal of a battery or the positive electrode of an electrolysis cell.

anodising: a process that uses the effect of electrolysis to make a surface corrosion-resistant.

antacid: a common name for any compound that reacts with stomach acid to neutralise it.

antioxidant: a substance that prevents oxidation of some other substance.

aqueous: a solid dissolved in water. Usually used as "aqueous solution".

atom: the smallest particle of an element.

atomic number: the number of electrons or the number of protons in an atom.

atomised: broken up into a very fine mist. The term is used in connection with sprays and engine fuel systems.

aurora: the "northern lights" and "southern lights" that show as coloured bands of light in the night sky at high latitudes. They are associated with the way cosmic rays interact with oxygen and nitrogen in the air.

basalt: an igneous rock with a low proportion of silica (usually below 55%). It has microscopically small crystals.

base: a compound that may be soapy to the touch and that can react with an acid in water to form a salt and water.

battery: a series of electrochemical cells.

bauxite: an ore of aluminium, of which about half is aluminium oxide.

becquerel: a unit of radiation equal to one nuclear disintegration per second.

beta particle: a form of radiation in which electrons are emitted from an atom as the nucleus breaks down.

bleach: a substance that removes stains from materials either by oxidising or reducing the staining compound.

boiling point: the temperature at which a liquid boils, changing from a liquid to a gas.

bond: chemical bonding is either a transfer or sharing of electrons by two or more atoms. There are a number of types of chemical bond, some very strong (such as covalent bonds), others weak (such as hydrogen bonds). Chemical bonds form because the linked molecule is more stable than the unlinked atoms from which it formed. For example, the hydrogen molecule (H_2) is more stable than single atoms of hydrogen, which is why hydrogen gas is always found as molecules of two hydrogen atoms.

brass: a metal alloy principally of copper and zinc.

brazing: a form of soldering, in which brass is used as the joining metal.

brine: a solution of salt (sodium chloride) in water.

bronze: an alloy principally of copper and tin.

buffer: a chemistry term meaning a mixture of substances in solution that resists a change in the acidity or alkalinity of the solution.

capillary action: the tendency of a liquid to be sucked into small spaces, such as between objects and through narrow-pore tubes. The force to do this comes from surface tension.

catalyst: a substance that speeds up a chemical reaction but itself remains unaltered at the end of the reaction.

cathode: the positive terminal of a battery or the negative electrode of an electrolysis cell.

cathodic protection: the technique of making the object that is to be protected from corrosion into the cathode of a cell. For example, a material, such as steel, is protected by coupling it with a more reactive metal, such as magnesium. Steel forms the cathode and magnesium the anode. Zinc protects steel in the same way.

cation: a positively charged atom or group of atoms.

caustic: a substance that can cause burns if it touches the skin.

cell: a vessel containing two electrodes and an electrolyte that can act as an electrical conductor.

ceramic: a material based on clay minerals, which has been heated so that it has chemically hardened.

chalk: a pure form of calcium carbonate made of the crushed bodies of microscopic sea creatures, such as plankton and algae.

change of state: a change between one of the three states of matter, solid, liquid and gas.

chlorination: adding chlorine to a substance.

cladding: a surface sheet of material designed to protect other materials from corrosion.

clay: a microscopically small plate-like mineral that makes up the bulk of many soils. It has a sticky feel when wet.

combustion: the special case of oxidisation of a substance where a considerable amount of heat and usually light are given out. Combustion is often referred to as "burning".

compound: a chemical consisting of two or more elements chemically bonded together. Calcium atoms can combine with carbon atoms and oxygen atoms to make calcium carbonate, a compound of all three atoms.

condensation nuclei: microscopic particles of dust, salt and other materials suspended in the air, which attract water molecules.

conduction: (i) the exchange of heat (heat conduction) by contact with another object or (ii) allowing the flow of electrons (electrical conduction).

convection: the exchange of heat energy with the surroundings produced by the flow of a fluid due to being heated or cooled.

corrosion: the *slow* decay of a substance resulting from contact with gases and liquids in the environment. The term is often applied to metals. Rust is the corrosion of iron.

corrosive: a substance, either an acid or an alkali, that *rapidly* attacks a wide range of other substances.

cosmic rays: particles that fly through space and bombard all atoms on the Earth's surface. When they interact with the atmosphere they produce showers of secondary particles.

covalent bond: the most common form of strong chemical bonding, which occurs when two atoms *share* electrons.

cracking: breaking down complex molecules into simpler components. It is a term particularly used in oil refining.

crude oil: a chemical mixture of petroleum liquids. Crude oil forms the raw material for an oil refinery.

crystal: a substance that has grown freely so that it can develop external faces. Compare with crystalline, where the atoms are not free to form individual crystals and amorphous where the atoms are arranged irregularly.

crystalline: the organisation of atoms into a rigid "honeycomb-like" pattern without distinct crystal faces.

crystal systems: seven patterns or systems into which all of the world's crystals can be grouped. They are: cubic, hexagonal, rhombohedral, tetragonal, orthorhombic, monoclinic and triclinic.

cubic crystal system: groupings of crystals that look like cubes.

curie: a unit of radiation. The amount of radiation emitted by 1 g of radium each second. (The curie is equal to 37 billion becquerels.)

current: an electric current is produced by a flow of electrons through a conducting solid or ions through a conducting liquid.

decay (radioactive decay): the way that a radioactive element changes into another element decause of loss of mass through radiation. For example uranium decays (changes) to lead.

decompose: to break down a substance (for example by heat or with the aid of a catalyst) into simpler components. In such a chemical reaction only one substance is involved.

dehydration: the removal of water from a substance by heating it, placing it in a dry atmosphere, or through the action of a drying agent.

density: the mass per unit volume (e.g. g/cc).

desertification: a process whereby a soil is allowed to become degraded to a state in which crops can no longer grow, i.e. desert-like. Chemical desertification is usually the result of contamination with halides because of poor irrigation practices.

detergent: a petroleum-based chemical that removes dirt.

diaphragm: a semipermeable membrane – a kind of ultra-fine mesh filter – that will allow only small ions to pass through. It is used in the electrolysis of brine.

diffusion: the slow mixing of one substance with another until the two substances are evenly mixed.

digestive tract: the system of the body that forms the pathway for food and its waste products. It begins at the mouth and includes the stomach and the intestines.

dilute acid: an acid whose concentration has been reduced by a large proportion of water.

diode: a semiconducting device that allows an electric current to flow in only one direction.

disinfectant: a chemical that kills bacteria and other microorganisms.

dissociate: to break apart. In the case of acids it means to break up forming hydrogen ions. This is an example of ionisation. Strong acids dissociate completely. Weak acids are not completely ionised and a solution of a weak acid has a relatively low concentration of hydrogen ions.

dissolve: to break down a substance in a solution without a resultant reaction.

distillation: the process of separating mixtures by condensing the vapours through cooling.

doping: adding metal atoms to a region of silicon to make it semiconducting.

dye: a coloured substance that will stick to another substance, so that both appear coloured.

electrode: a conductor that forms one terminal of a cell.

electrolysis: an electrical–chemical process that uses an electric current to cause the break up of a compound and the movement of metal ions in a solution. The process happens in many natural situations (as for example in rusting) and is also commonly used in industry for purifying (refining) metals or for plating metal objects with a fine, even metal coating.

electrolyte: a solution that conducts electricity.

electron: a tiny, negatively charged particle that is part of an atom. The flow of electrons through a solid material such as a wire produces an electric current.

electroplating: depositing a thin layer of a metal onto the surface of another substance using electrolysis.

element: a substance that cannot be decomposed into simpler substances by chemical means

emulsion: tiny droplets of one substance dispersed in another. A common oil in water emulsion is milk. The tiny droplets in an emulsion tend to come together, so another stabilising substance is often needed to wrap the particles of grease and oil in a stable coat. Soaps and detergents are such agents. Photographic film is an example of a solid emulsion.

endothermic reaction: a reaction that takes heat from the surroundings. The reaction of carbon monoxide with a metal oxide is an example.

enzyme: organic catalysts in the form of proteins in the body that speed up chemical reactions. Every living cell contains hundreds of enzymes, which ensure that the processes of life continue. Should enzymes be made inoperative, such as through mercury poisoning, then death follows.

ester: organic compounds, formed by the reaction of an alcohol with an acid, which often have a fruity taste.

evaporation: the change of state of a liquid to a gas. Evaporation happens below the boiling point and is used as a method of separating out the materials in a solution.

exothermic reaction: a reaction that gives heat to the surroundings. Many oxidation reactions, for example, give out heat.

explosive: a substance which, when a shock is applied to it, decomposes very rapidly, releasing a very large amount of heat and creating a large volume of gases as a shock wave.

extrusion: forming a shape by pushing it through a die. For example, toothpaste is extruded through the cap (die) of the toothpaste tube.

fallout: radioactive particles that reach the ground from radioactive materials in the atmosphere.

fat: semi-solid energy-rich compounds derived from plants or animals and which are made of carbon, hydrogen and oxygen. Scientists call these esters.

feldspar: a mineral consisting of sheets of aluminium silicate. This is the mineral from which the clay in soils is made.

fertile: able to provide the nutrients needed for unrestricted plant growth.

filtration: the separation of a liquid from a solid using a membrane with small holes.

fission: the breakdown of the structure of an atom, popularly called "splitting the atom" because the atom is split into approximately two other nuclei. This is different from, for example, the small change that happens when radioactivity is emitted.

fixation of nitrogen: the processes that natural organisms, such as bacteria, use to turn the nitrogen of the air into ammonium compounds.

fixing: making solid and liquid nitrogen-containing compounds from nitrogen gas. The compounds that are formed can be used as fertilisers.

fluid: able to flow; either a liquid or a gas.

fluorescent: a substance that gives out visible light when struck by invisible waves such as ultraviolet rays.

flux: a material used to make it easier for a liquid to flow. A flux dissolves metal oxides and so prevents a metal from oxidising while being heated.

foam: a substance that is sufficiently gelatinous to be able to contain bubbles of gas. The gas bulks up the substance, making it behave as though it were semi-rigid.

fossil fuels: hydrocarbon compounds that have been formed from buried plant and animal remains. High pressures and temperatures lasting over millions of years are required. The fossil fuels are coal, oil and natural gas.

fraction: a group of similar components of a mixture. In the petroleum industry the light fractions of crude oil are those with the smallest molecules, while the medium and heavy fractions have larger molecules.

free radical: a very reactive atom or group with a "spare" electron.

freezing point: the temperature at which a substance changes from a liquid to a solid. It is the same temperature as the melting point.

fuel: a concentrated form of chemical energy. The main sources of fuels (called fossil fuels because they were formed by geological processes) are coal, crude oil and natural gas. Products include methane, propane and gasoline. The fuel for stars and space vehicles is hydrogen.

fuel rods: rods of uranium or other radioactive material used as a fuel in nuclear power stations.

fuming: an unstable liquid that gives off a gas. Very concentrated acid solutions are often fuming solutions.

fungicide: any chemical that is designed to kill fungi and control the spread of fungal spores.

fusion: combining atoms to form a heavier atom.

galvanising: applying a thin zinc coating to protect another metal.

gamma rays: waves of radiation produced as the nucleus of a radioactive element rearranges itself into a tighter cluster of protons and neutrons. Gamma rays carry enough energy to damage living cells.

gangue: the unwanted material in an ore.

gas: a form of matter in which the molecules form no definite shape and are free to move about to fill any vessel they are put in.

gelatinous: a term meaning made with water. Because a gelatinous precipitate is mostly water, it is of a similar density to water and will float or lie suspended in the liquid.

gelling agent: a semi-solid jelly-like substance.

gemstone: a wide range of minerals valued by people, both as crystals (such as emerald) and as decorative stones (such as agate). There is no single chemical formula for a gemstone.

glass: a transparent silicate without any crystal growth. It has a glassy lustre and breaks with a curved fracture. Note that some minerals have all these features and are therefore natural glasses. Household glass is a synthetic silicate.

glucose: the most common of the natural sugars. It occurs as the polymer known as cellulose, the fibre in plants. Starch is also a form of glucose. The breakdown of glucose provides the energy that animals need for life.

granite: an igneous rock with a high proportion of silica (usually over 65%). It has well-developed large crystals. The largest pink, grey or white crystals are feldspar.

Greenhouse Effect: an increase of the global air temperature as a result of heat released from burning fossil fuels being absorbed by carbon dioxide in the atmosphere.

gypsum: the name for calcium sulphate. It is commonly found as Plaster of Paris and wallboards.

half-life: the time it takes for the radiation coming from a sample of a radioactive element to decrease by half.

halide: a salt of one of the halogens (fluorine, chlorine, bromine and iodine).

halite: the mineral made of sodium chloride.

halogen: one of a group of elements including chlorine, bromine, iodine and fluorine.

heat-producing: see exothermic reaction.

high explosive: a form of explosive that will only work when it receives a shock from another explosive. High explosives are much more powerful than ordinary explosives. Gunpowder is not a high explosive.

hydrate: a solid compound in crystalline form that contains molecular water. Hydrates commonly form when a solution of a soluble salt is evaporated. The water that forms part of a hydrate crystal is known as the "water of crystallization". It can usually be removed by heating, leaving an anhydrous salt.

hydration: the absorption of water by a substance. Hydrated materials are not "wet" but remain firm, apparently dry, solids. In some cases, hydration makes the substance change colour, in many other cases there is no colour change, simply a change in volume.

hydrocarbon: a compound in which only hydrogen and carbon atoms are present. Most fuels are hydrocarbons, as is the simple plastic polyethene (known as polythene).

hydrogen bond: a type of attractive force that holds one molecule to another. It is one of the weaker forms of intermolecular attractive force.

hydrothermal: a process in which hot water is involved. It is usually used in the context of rock formation because hot water and other fluids sent outwards from liquid magmas are important carriers of metals and the minerals that form gemstones.

igneous rock: a rock that has solidified from molten rock, either volcanic lava on the Earth's surface or magma deep underground. In either case the rock develops a network of interlocking crystals.

incendiary: a substance designed to cause burning.

indicator: a substance or mixture of substances that change colour with acidity or alkalinity.

inert: nonreactive.

infra-red radiation: a form of light radiation where the wavelength of the waves is slightly longer than visible light. Most heat radiation is in the infra-red band.

insoluble: a substance that will not dissolve.

ion: an atom, or group of atoms, that has gained or lost one or more electrons and so developed an electrical charge. Ions behave differently from electrically neutral atoms and molecules. They can move in an electric field, and they can also bind strongly to solvent molecules such as water. Positively charged ions are called cations; negatively charged ions are called anions. Ions carry electrical current through solutions.

ionic bond: the form of bonding that occurs between two ions when the ions have opposite charges. Sodium cations bond with chloride anions to form common salt (NaCl) when a salty solution is evaporated. Ionic bonds are strong bonds except in the presence of a solvent.

ionise: to break up neutral molecules into oppositely charged ions or to convert atoms into ions by the loss of electrons.

ionisation: a process that creates ions.

irrigation: the application of water to fields to help plants grow during times when natural rainfall is sparse.

isotope: atoms that have the same number of protons in their nucleus, but which have different masses; for example, carbon-12 and carbon-14.

latent heat: the amount of heat that is absorbed or released during the process of changing state between gas, liquid or solid. For example, heat is absorbed when a substance melts and it is released again when the substance solidifies.

latex: (the Latin word for "liquid") a suspension of small polymer particles in water. The rubber that flows from a rubber tree is a natural latex. Some synthetic polymers are made as latexes, allowing polymerisation to take place in water.

lava: the material that flows from a volcano.

limestone: a form of calcium carbonate rock that is often formed of lime mud. Most limestones are light grey and have abundant fossils.

liquid: a form of matter that has a fixed volume but no fixed shape.

lode: a deposit in which a number of veins of a metal found close together.

lustre: the shininess of a substance.

magma: the molten rock that forms a balloon-shaped chamber in the rock below a volcano. It is fed by rock moving upwards from below the crust.

marble: a form of limestone that has been "baked" while deep inside mountains. This has caused the limestone to melt and reform into small interlocking crystals, making marble harder than limestone.

mass: the amount of matter in an object. In everyday use, the word weight is often used to mean mass.

melting point: the temperature at which a substance changes state from a solid to a liquid. It is the same as freezing point.

membrane: a thin flexible sheet. A semipermeable membrane has microscopic holes of a size that will selectively allow some ions and molecules to pass through but hold others back. It thus acts as a kind of sieve.

meniscus: the curved surface of a liquid that forms when it rises in a small bore, or capillary tube. The meniscus is convex (bulges upwards) for mercury and is concave (sags downwards) for water.

metal: a substance with a lustre, the ability to conduct heat and electricity and which is not brittle.

metallic bonding: a kind of bonding in which atoms reside in a "sea" of mobile electrons. This type of bonding allows metals to be good conductors and means that they are not brittle

metamorphic rock: formed either from igneous or sedimentary rocks, by heat and or pressure. Metamorphic rocks form deep inside mountains during periods of mountain building. They result from the remelting of rocks during which process crystals are able to grow. Metamorphic rocks often show signs of banding and partial melting.

micronutrient: an element that the body requires in small amounts. Another term is trace element.

mineral: a solid substance made of just one element or chemical compound. Calcite is a mineral because it consists only of calcium carbonate, halite is a mineral because it contains only sodium chloride, quartz is a mineral because it consists of only silicon dioxide.

mineral acid: an acid that does not contain carbon and that attacks minerals. Hydrochloric, sulphuric and nitric acids are the main mineral acids.

mineral-laden: a solution close to saturation.

mixture: a material that can be separated out into two or more substances using physical means.

molecule: a group of two or more atoms held together by chemical bonds.

monoclinic system: a grouping of crystals that look like double-ended chisel blades.

monomer: a building block of a larger chain molecule ("mono" means one, "mer" means part).

mordant: any chemical that allows dyes to stick to other substances.

native metal: a pure form of a metal, not combined as a compound. Native metal is more common in poorly reactive elements than in those that are very reactive.

neutralisation: the reaction of acids and bases to produce a salt and water. The reaction causes hydrogen from the acid and hydroxide from the base to be changed to water. For example, hydrochloric acid reacts with sodium hydroxide to form common salt and water. The term is more generally used for any reaction where the pH changes towards 7.0, which is the pH of a neutral solution.

neutron: a particle inside the nucleus of an atom that is neutral and has no charge.

noncombustible: a substance that will not burn.

noble metal: silver, gold, platinum, and mercury. These are the least reactive metals.

nuclear energy: the heat energy produced as part of the changes that take place in the core, or nucleus, of an element's atoms.

nuclear reactions: reactions that occur in the core, or nucleus of an atom.

nutrients: soluble ions that are essential to life.

octane: one of the substances contained in fuel.

ore: a rock containing enough of a useful substance to make mining it worthwhile.

organic acid: an acid containing carbon and hydrogen.

organic substance: a substance that contains carbon.

osmosis: a process where molecules of a liquid solvent move through a membrane (filter) from a region of low concentration to a region of high concentration of solute.

oxidation: a reaction in which the oxidising agent removes electrons. (Note that oxidising agents do not have to contain oxygen.)

oxide: a compound that includes oxygen and one other element.

oxidise: the process of gaining oxygen. This can be part of a controlled chemical reaction, or it can be the result of exposing a substance to the air, where oxidation (a form of corrosion) will occur slowly, perhaps over months or years.

oxidising agent: a substance that removes electrons from another substance (and therefore is itself reduced).

ozone: a form of oxygen whose molecules contain three atoms of oxygen. Ozone is regarded as a beneficial gas when high in the atmosphere because it blocks ultraviolet rays. It is a harmful gas when breathed in, so low level ozone, which is produced as part of city smog, is regarded as a form of pollution. The ozone layer is the uppermost part of the stratosphere.

pan: the name given to a shallow pond of liquid. Pans are mainly used for separating solutions by evaporation.

patina: a surface coating that develops on metals and protects them from further corrosion.

percolate: to move slowly through the pores of a rock.

period: a row in the Periodic Table.

Periodic Table: a chart organising elements by atomic number and chemical properties into groups and periods.

pesticide: any chemical that is designed to control pests (unwanted organisms) that are harmful to plants or animals.

petroleum: a natural mixture of a range of gases, liquids and solids derived from the decomposed remains of plants and animals.

pH: a measure of the hydrogen ion concentration in a liquid. Neutral is pH 7.0; numbers greater than this are alkaline, smaller numbers are acidic.

phosphor: any material that glows when energized by ultraviolet or electron beams such as in fluorescent tubes and cathode ray tubes. Phosphors, such as phosphorus, emit light after the source of excitation is cut off. This is why they glow in the dark. By contrast, fluorescors, such as fluorite, emit light only while they are being excited by ultraviolet light or an electron beam.

photon: a parcel of light energy.

photosynthesis: the process by which plants use the energy of the Sun to make the compounds they need for life. In photosynthesis, six molecules of carbon dioxide from the air combine with six molecules of water, forming one molecule of glucose (sugar) and releasing six molecules of oxygen back into the atmosphere.

pigment: any solid material used to give a liquid a colour.

placer deposit: a kind of ore body made of a sediment that contains fragments of gold ore eroded from a mother lode and transported by rivers and/or ocean currents.

plastic (material): a carbon-based material consisting of long chains (polymers) of simple molecules. The word plastic is commonly restricted to synthetic polymers.

plastic (property): a material is plastic if it can be made to change shape easily. Plastic materials will remain in the new shape. (Compare with elastic, a property where a material goes back to its original shape.)

plating: adding a thin coat of one material to another to make it resistant to corrosion.

playa: a dried-up lake bed that is covered with salt deposits. From the Spanish word for beach.

poison gas: a form of gas that is used intentionally to produce widespread injury and death. (Many gases are poisonous, which is why many chemical reactions are performed in laboratory fume chambers, but they are a byproduct of a reaction and not intended to cause harm.)

polymer: a compound that is made of long chains by combining molecules (called monomers) as repeating units. ("Poly" means many, "mer" means part).

polymerisation: a chemical reaction in which large numbers of similar molecules arrange themselves into large molecules, usually long chains. This process usually happens when there is a suitable catalyst present. For example, ethene reacts to form polythene in the presence of certain catalysts.

porous: a material containing many small holes or cracks. Quite often the pores are connected, and liquids, such as water or oil, can move through them.

precious metal: silver, gold, platinum, iridium, and palladium. Each is prized for its rarity. This category is the equivalent of precious stones, or gemstones, for minerals.

precipitate: tiny solid particles formed as a result of a chemical reaction between two liquids or gases.

preservative: a substance that prevents the natural organic decay processes from occurring. Many substances can be used safely for this purpose, including sulphites and nitrogen gas.

product: a substance produced by a chemical reaction.

protein: molecules that help to build tissue and bone and therefore make new body cells. Proteins contain amino acids.

proton: a positively charged particle in the nucleus of an atom that balances out the charge of the surrounding electrons

pyrite: "mineral of fire". This name comes from the fact that pyrite (iron sulphide) will give off sparks if struck with a stone.

pyrometallurgy: refining a metal from its ore using heat. A blast furnace or smelter is the main equipment used.

radiation: the exchange of energy with the surroundings through the transmission of waves or particles of energy. Radiation is a form of energy transfer that can happen through space; no intervening medium is required (as would be the case for conduction and convection).

radioactive: a material that emits radiation or particles from the nucleus of its atoms.

radioactive decay: a change in a radioactive element due to loss of mass through radiation. For example uranium decays (changes) to lead.

radioisotope: a shortened version of the phrase radioactive isotope.

radiotracer: a radioactive isotope that is added to a stable, nonradioactive material in order to trace how it moves and its concentration.

reaction: the recombination of two substances using parts of each substance to produce new substances.

reactivity: the tendency of a substance to react with other substances. The term is most widely used in comparing the reactivity of metals. Metals are arranged in a reactivity series.

reagent: a starting material for a reaction.

recycling: the reuse of a material to save the time and energy required to extract new material from the Earth and to conserve non-renewable resources.

redox reaction: a reaction that involves reduction and oxidation.

reducing agent: a substance that gives electrons to another substance. Carbon monoxide is a reducing agent when passed over copper oxide, turning it to copper and producing carbon dioxide gas. Similarly, iron oxide is reduced to iron in a blast furnace. Sulphur dioxide is a reducing agent, used for bleaching bread.

reduction: the removal of oxygen from a substance. See also: oxidation.

refining: separating a mixture into the simpler substances of which it is made. In the case of a rock, it means the extraction of the metal that is mixed up in the rock. In the case of oil it means separating out the fractions of which it is made.

refractive index: the property of a transparent material that controls the angle at which total internal reflection will occur. The greater the refractive index, the more reflective the material will be.

resin: natural or synthetic polymers that can be moulded into solid objects or spun into thread.

rust: the corrosion of iron and steel.

saline: a solution in which most of the dissolved matter is sodium chloride (common salt).

salinisation: the concentration of salts, especially sodium chloride, in the upper layers of a soil due to poor methods of irrigation.

salts: compounds, often involving a metal, that are the reaction products of acids and bases. (Note "salt" is also the common word for sodium chloride, common salt or table salt.)

saponification: the term for a reaction between a fat and a base that produces a soap.

saturated: a state where a liquid can hold no more of a substance. If any more of the substance is added, it will not dissolve.

saturated solution: a solution that holds the maximum possible amount of dissolved material. The amount of material in solution varies with the temperature; cold solutions can hold less dissolved solid material than hot solutions. Gases are more soluble in cold liquids than hot liquids.

sediment: material that settles out at the bottom of a liquid when it is still.

semiconductor: a material of intermediate conductivity. Semiconductor devices often use silicon when they are made as part of diodes, transistors or integrated circuits.

semipermeable membrane: a thin (membrane) of material that acts as a fine sieve, allowing small molecules to pass, but holding large molecules back.

silicate: a compound containing silicon and oxygen (known as silica).

sintering: a process that happens at moderately high temperatures in some compounds. Grains begin to fuse together even through they do not melt. The most widespread example of sintering happens during the firing of clays to make ceramics.

slag: a mixture of substances that are waste products of a furnace. Most slags are composed mainly of silicates.

smelting: roasting a substance in order to extract the metal contained in it.

smog: a mixture of smoke and fog. The term is used to describe city fogs in which there is a large proportion of particulate matter (tiny pieces of carbon from exhausts) and also a high concentration of sulphur and nitrogen gases and probably ozone.

soldering: joining together two pieces of metal using solder, an alloy with a low melting point.

solid: a form of matter where a substance has a definite shape.

soluble: a substance that will readily dissolve in a solvent.

solute: the substance that dissolves in a solution (e.g. sodium chloride in salt water).

solution: a mixture of a liquid and at least one other substance (e.g. salt water). Mixtures can be separated out by physical means, for example by evaporation and cooling.

solvent: the main substance in a solution (e.g. water in salt water).

spontaneous combustion: the effect of a very reactive material beginning to oxidise very quickly and bursting into flame.

stable: able to exist without changing into another substance.

stratosphere: the part of the Earth's atmosphere that lies immediately above the region in which clouds form. It occurs between 12 and 50 km above the Earth's surface.

strong acid: an acid that has completely dissociated (ionised) in water. Mineral acids are strong acids.

sublimation: the change of a substance from solid to gas, or vica versa, without going through a liquid phase.

substance: a type of material, including mixtures.

sulphate: a compound that includes sulphur and oxygen, for example, calcium sulphate or gypsum.

sulphide: a sulphur compound that contains no oxygen.

sulphite: a sulphur compound that contains less oxygen than a sulphate.

surface tension: the force that operates on the surface of a liquid, which makes it act as though it were covered with an invisible elastic film.

suspension: tiny particles suspended in a liquid.

synthetic: does not occur naturally, but has to be manufactured.

tarnish: a coating that develops as a result of the reaction between a metal and substances in the air. The most common form of tarnishing is a very thin transparent oxide coating.

thermonuclear reactions: reactions that occur within atoms due to fusion, releasing an immensely concentrated amount of energy.

thermoplastic: a plastic that will soften, can repeatedly be moulded it into shape on heating and will set into the moulded shape as it cools.

thermoset: a plastic that will set into a moulded shape as it cools, but which cannot be made soft by reheating.

titration: a process of dripping one liquid into another in order to find out the amount needed to cause a neutral solution. An indicator is used to signal change.

toxic: poisonous enough to cause death.

translucent: almost transparent.

transmutation: the change of one element into another.

vapour: the gaseous form of a substance that is normally a liquid. For example, water vapour is the gaseous form of liquid water.

vein: a mineral deposit different from, and usually cutting across, the surrounding rocks. Most mineral and metal-bearing veins are deposits filling fractures. The veins were filled by hot, mineral-rich waters rising upwards from liquid volcanic magma. They are important sources of many metals, such as silver and gold, and also minerals such as gemstones. Veins are usually narrow, and were best suited to hand-mining. They are less exploited in the modern machine age.

viscous: slow moving, syrupy. A liquid that has a low viscosity is said to be mobile.

vitreous: glass-like.

volatile: readily forms a gas.

vulcanisation: forming cross-links between polymer chains to increase the strength of the whole polymer. Rubbers are vulcanised using sulphur when making tyres and other strong materials.

weak acid: an acid that has only partly dissociated (ionised) in water. Most organic acids are weak acids.

weather: a term used by Earth scientists and derived from "weathering", meaning to react with water and gases of the environment.

weathering: the slow natural processes that break down rocks and reduce them to small fragments either by mechanical or chemical means.

welding: fusing two pieces of metal together using heat.

X-rays: a form of very short wave radiation.

Index